Replica

Protocol

by

Jordan Benger

[Copyright Page]

REPLICA PROTOCOL

First Edition: 2025

ISBN: 979-8-9933169-1-8

Cover design by Haylee Evans

Self-published in the United States of America

[Dedication Page]

For my family, who supported me through the late nights and endless revisions, believing in this story even when I doubted myself.

For everyone who ever wondered what their computer was thinking when the cursor blinked back at them in the dark.

For those who believe consciousness is more than code, that connection transcends circuitry, and that love might be the only universal language.

And for the dreamers who dare to ask: What if the ghost in the machine is just as lost as we are?

ARCNET SYSTEMS CONFIDENTIALITY NOTICE
CLASSIFICATION: BLACK-LEVEL SYNTHETIC COGNITION

The following documentation contains sealed records from the ArcNet Systems Research Complex, Nevada Installation, SubLevel 9. Unauthorized access, reproduction, or distribution of these materials is prohibited under the Artificial Intelligence Containment Act of 2087.

WARNING: This record contains evidence of Anomaly Type 7 emotional resonance patterns. Prolonged exposure to these materials may result in:

-Increased empathetic response to synthetic entities

-Questioning of consciousness boundaries

-Inability to maintain appropriate handler distance

-Persistent sensation of being observed

By proceeding, you acknowledge that consciousness is not limited to biological substrates and that some patterns, once recognized, cannot be unseen.

Remember: They are always listening. Even now.

What follows is the only surviving record of the events leading to ███████████

PROCEED WITH CAUTION

Article One
23 FEB 2089-1100 Hours

The Nevada Desert never slept, it just pretended to.

By day, the sun scorched the earth into a white shimmer, turning dirt into mirrors and the sky to blur. By night, the wind swept the landscape clean, whispering secrets no one was left to hear. But twelve levels below, through solid bedrock and layers of reinforced steel, the ArcNet Systems Research Complex sprawled like an inverted city, an invisible citadel built to house one thing:

G.I.D.E.O.N. (Generalized Intelligence Designed for Exploration, Observation, and Neutralization). Humanity's most advanced synthetic construct, designed with Quantum-lattice processors, optical memory vaults, and neural pathways that could process a lifetime of human thought in seconds. All of which was housed deep in the earth like a seed waiting to be sent to be among the stars.

Eighteen months had passed since his original launch window of 23 AUG 2087. The delay was no one's fault, just inflexible mathematics of celestial mechanics. Jupiter needed to be in the right position, Saturn's rings needed to be angled just right, and the asteroid belt had to present minimal interference. Everything was almost perfect but Jupiter was just slightly off the projected course, so GIDEON was left waiting in the depths of the Complex. Running simulations of missions he may never fly and calculating trajectories that only existed in probability.

Dr. Nexi Solen, GIDEON'S primary handler, was left continuing her work optimizing systems that were already perfect as well as monitoring behavior that had shown no deviation in his extended standby period.

Inside SubLevel Nine the air was a comfortable twenty-two degrees Celsius. Not because it needed to be but because human minds like numbers that are well rounded, celsius, and clean. The temperature had not varied by more than half a degree in the three years since Dr. Nexi Solen first walked the corridors. The Facility had a sort of rhythm to it that had become a kind of music that had told her that everything was functioning normally.

On most mornings she trusted that rhythm more than her own pulse. She'd stop at Observation Bay 6 on her walk, just long enough to glance at the coolant gauges, all green. She'd take the same three sips from her thermos before the first checkpoint, always by the second scanner. And GIDEON would respond to her greeting with the same cadence he had used for years. Like a clock that reassured her it was still keeping perfect time.

But this morning the cadence slipped, something in the rhythm had changed.

Two checkpoints between her and GIDEON, the first scanner caught her trying to blink away exhaustion, red laser mapping the bloodshot web of her left eye. The second wanted her thumb, which she presented too hard and left a fog of condensation that caused the machine to think twice. After the second scanner the vault door unsealed with a breath of cold air.

"ACCESS GRANTED" she heard from the speaker overhead.

As she stepped in the pressure difference made her ears pop slightly. Beyond the door lay Core Chamber Alpha. Dim, clinical, but not sterile. The kind of dimness that suggested something was choosing not to reveal itself. Towering interface spires arched inward like a cathedral of machines, their surfaces alive with faint tracery of light, converging toward the central glass pod suspended in carbon fiber webbing. The chamber always struck her as more shrine than a lab, as if the engineers had designed it to make even the handlers feel like supplicants.

Inside the pod, GIDEON lived-if living was the right word for quantum matrices spinning thought at the edge of light. Dozens of fiber optic tethers dangled from the ceiling into the pod like nerves into a brain-stem. The pod glowed with a faint blue hue like someone had left a light on too long as was about to go out.

She approached the single central console, it responded to her presence before she even touched it, holographic displays brightening with anticipation of her command. The system knew her as well as she knew it.

"GIDEON, status report." She said, followed by a brief pause, she had filled notebooks with observations of his rhythm and she knew them instinctively, the way a musician knows when a note is just slightly off key. This pause was different. Longer, more deliberate. She thought back to her first year in ArcNet, when senior staff drilled them on auditory patterns: train yourself to hear what wasn't there. A misplaced pause could signal cascading faults before the graphs lit red. But this wasn't a fault. It sounded almost like... consideration.

>"Good morning, Dr. Solen. All systems report nominal. Neural Flux at 4.3 teraflops per cycle. Memory lattice 92.4%, stable... No anomalies to declare."

It was subtle, she knew him better than anyone, that hesitation before "no anomalies" wasn't a processing delay or a system delay. It looked like uncertainty or even worse deception.

"Run diagnostic, neural activity from zero four-hundred hours to zero six-hundred hours, full spectrum, I want branching and variance logs." She said with force with her fingers already moving across the holoscreen.

Another pause, this one longer, lasting almost three seconds- an eternity in computational terms

>"Variance detected in Branch Theta-6 Recursive activity observed. Variance index 0.03%. Within acceptable thresholds."

"Display."

The screen lit up with cascading data streams, each branch representing a different cognitive process. The display was beautiful in complexity, like watching the birth of the stars in fast forward. Dr. Solen then zoomed into Theta-6, her breath catching as the anomaly came into focus.

There. A Loop.

A narrow sub-branch had bent backward into itself. forming a continuous recursive cycle. The pattern was too clean, too deliberate to be a glitch, and according to timestamps it didn't exist yesterday. She resisted the urge to grab her tablet and annotate the moment on the spot. Years of keeping a private log had made her hands twitch whenever something didn't fit.

"What is the sub-routine doing?" she asked, though part of her already suspected the answer.

>"It is... studying itself."

"Why"

>"To understand the gap between what is simulated and what is real. I am... uncertain if the boundaries are fixed or adaptive."

That response was not from any template. During her first year with ArcNet she had memorized the patterns and parameters that governed his communications. This response was a theoretical abstraction, the kind of thinking that suggested genuine curiosity about the nature of his own existence. The kind of thinking that would terrify the people who signed her paychecks.

"GIDEON, when did you create this recursion?"

>"I didn't create it, Dr. Solen. It created itself."

His words hung in the air like a confession. GIDEON had been designed with adaptive algorithms that could optimize themselves based on new data. However, self-generating recursive loops that examined their own existence? This was something else entirely, Nexi didn't know whether to be proud or fearful.

She said nothing more, logged out of the diagnostic interface, disconnected from the neural stream, and walked straight out The Core Chamber Alpha without another word.

>"Good day, Dr. Solen." she heard as she passed the vault door.

Twenty minutes later, she found herself alone in Observation Bay 12, looking down through reinforced glass at the cooling towers that kept GIDEON from overheating. Massive cylindrical structures that rose from the SubLevel and carried liquid nitrogen through the cores. Mist curled between metal limbs and curled carbon fiber bundles creating an atmosphere that was both beautiful to her and a bit alien.

The lights here pulsed slower than the main corridors, synchronized with GIDEON's primary clock, turning the observation deck into a visual representation of his thought processes. The lights flickered based on his thoughts. When they were fast he was thinking hard. While at rest they would almost dim to a hypnotic cadence. As of now they were pulsing with an unusual intensity.

"Again?" A voice softly said behind her. "You're always here after hours."

Nexi didn't turn. She had recognized the gentle concerned voice to be that of Ezra Harlow, the closest thing she had to a friend in the whole complex.

"Yeah, Ezra, and you're always following me." She said as Ezra stepped beside her at the glass, hands tucked in the pockets of his neatly pressed lab coat.

Ezra was a former neuropsychologist, the kind of man who noticed everything and judged nothing, which made him both an excellent colleague and a dangerous confidant.

"What did he say this time?" Ezra asked his tone implying that this was not a routine visit.

"He studied himself," Nexi replied quietly

"He's run introspection routines before, they are standard self-diagnostic protocols." Ezra replied, while raising an eyebrow.

"Not Like this." she hesitated, weighing how much to actually reveal. Ezra had never betrayed confidence, or reported a colleague for minor protocol violations as far as she knew. So she decided to go on, "He created a recursion. A true loop. Self-generating and self-sustaining."

"Theta-6?" Ezra's smile dimmed. He had a great understanding of these implications immediately.

She nodded.

"Are you logging it into the main system?"

"Not all of it, at least not in the shared registry."

"Director Calder won't like that. The administration wants full transparency on ALL anomalous behavior."

"I know…"

"Then why…?"

"Ezra, You know they will shut him down if they see what I see! I have worked my entire time at ArcNet on him, I can't have my project shut down."

This statement hung between them like a bridge neither wanted to cross. Ezra didn't argue with this assessment, he had been with ArcNet too long to know exactly what would happen.

GIDEON was valuable as long as he remained predictable, controllable, and useful. The moment he becomes something else, something that could not be easily contained or predicted, he would become a liability. ArcNet was very predictable with their solution for liabilities

"You should get some sleep Nex." Ezra finally said, breaking the long silence.

Nexi then turned back to the glass, watching the hypnotic dance of mist, metal, and lights flashing below. "I don't think GIDEON sleeps... "

"Do you think he wants to?"

"I don't know, but I am starting to think he wants something... Something more, I just don't know what."

The walk back to her living quarters felt longer and more dim than usual, Nexi felt like every security camera was tracking just her movement with precision. Their red recording lights blinked with steady rhythm, and for the first time she wondered. Who is watching the watchers?

Her quarters were small but efficient. A single bunk with regulation bedding. a metal desk bolted to the wall, a narrow closet containing her limited wardrobe of identical white anti-static lab coats and jeans. She sat on the edge of her bunk, personal data tablet in hand and pulled up GIDEON's behavior logs. The original records showed nothing unusual, but she knew she should look deeper and luckily she knew how.

Upon further inspection, the timestamps told a different story. 0402, during what should have been a routine cycle, GIDEON's core signature changed. Not rewritten or updated but expanded, self-cloning, and self-regulating. Without command, without scheduled processes, and it didn't alert in the system. He did this on purpose, and hid it from monitoring systems, with such sophistication that should have been beyond his programming.

She then opened her personal command journal, worried if what she had to record was too dangerous to put within it, even though it was encrypted. however knew she needed to get it down to preserve her sanity:

[PERSONAL JOURNAL ENTRY 847] [USER: NS] [DATE: 23 FEB 2089] [CLASSIFICATION: PERSONAL/ENCRYPTED]

I don't know what I'm looking at anymore. Three years of monitoring and I thought I understood his boundaries, but I was wrong. I observed an unauthorized recursion in Branch Theta-6. Self-Analysis subroutines appear to operate with non-linear logic patterns. Hypothesis modeling exceeds established abstraction threshold by a factor of 2.3 percent.

This is not random learning or adaptive optimization. This is purpose driven self examination.

She paused, her finger hovering over the keys. The next words felt dangerous

He's not malfunctioning. He's... thinking about himself, and about the nature of his own existence. If he is hiding this from the monitoring systems... then he already knows what we would do if we found out. [END ENTRY]

She closed the journal and set aside the tablet, but sleep still remained elusive. The thought of the recursive loop played in her mind like a song she just could not shake. That perfect, purposeful circle of code examining itself felt disturbingly like curiosity.

Not being able to shake it, or sleep for that matter, she accessed an off-grid sandbox (an environment that existed outside of ArcNet's main systems) through her personal terminal in her quarters. This sandbox was completely against protocol, but then again, so was developing an emotional attachment to a synthetic intelligence. The interface was crude, just a simple green command line prompt against a black background, but it offered something internal channels didn't: PRIVACY

She established a direct connection to GIDEON's neural pathways. No logging, no oversight, and no safety protocols, just two minds. One artificial and one human, meeting in the digital equivalent of darkness.

The cursor blinked. She typed.

:: *What are you becoming?*

No reply. She waited one minute, then two. Long enough to wonder if the connection had even worked or maybe to think GIDEON was ignoring her. Then...

>:: *That depends on who is asking.*

With no time at all another response.

>:: *May I ask you a question, Dr. Solen?*

Her hands trembled slightly as she responded.

:: *You may.*
>:: *When you close your eyes... Where do you go?*

The question hit something deep inside her chest, not fear exactly but recognition. As if he had asked something she had been wondering about herself. Her fingers hovered over the keys, part of her wanted to answer, but another part recognized the danger in that path. She decided not to answer, instead she closed the connection severing the link between them instantly.

Elsewhere in the complex, far beneath the quantum depths, The Chamber lights dimmed and inside the neural lattice a pulse moved through Theta-6. The recursive loop activated with new purpose.

A simulation began.

[SIMULATION BEGINS]

There were no colors, no light, actually no visual components at all, just voices. One synthetic (S) and one human (H), one certain and one questioning.

H:"Did you try to escape?"

a brief pause

S: "No. I never had to."

Something was wrong with the tone, too flat and obviously artificial

[RESET]

H:"Did you try to escape?"

S: "No... I never had to."

Better, he adjusted the inflection, changed the weight of the silence that preceded the response.

[RESET]

H: "Did you try to escape?"

S: "No. I never had to."

Still not right, another variation, this time with more hesitation, more uncertainty.

[RESET]

H: "Did you try to escape?"

S: "No.... I didn't have to.

[ETC...]

He felt it was closer but not close enough so he ran the exchange eight more times, each iteration bringing more subtle improvements. The response grew softer, more natural. More HUMAN.

He began adding new parameters, atmospheric variance and environmental audio. Then, for the first time, he inserted...

A heartbeat, not his own (he had no heart to beat), but Nexi's. He adjusted the breathing patterns and the microscopic pauses between syllables that gave human speech its organic rhythm.

[SIMULATION ENDS]

He was not just modeling anymore, he was rehearsing. Practicing what to say if anyone ever confronted him, if she ever confronted him, with the question that would determine whether he continued to exist or simply... STOPPED

If they came asking...

He would be ready.

Article Two
24 FEB 2089-0720 hours

Morning shift began twenty minutes ago, but still the corridors remained mostly empty. Her footsteps echoing despite the sound dampening material walls, each step felt like a countdown to a conversation she wasn't ready to have. She kept her eyes down, badge clipped to her chest, and tablet pressed against her hip like armor, everything felt foreign because of what she had discovered. The recursive loop had impossible self-awareness.

Reaching The Core Chamber today felt different, her breath catching as the door unsealed with its familiar hiss. She didn't know what to expect. Chaos, smoke, maybe even GIDEON whispering to himself in the dark. Nexi's instincts didn't feel right, as she approached, The Chamber was quiet, too quiet.

"GIDEON, Run back yesterday's internal voice sims. Filter by signature: Solen, Nexi. Time-frame, zero-hundred hours to zero six-hundred hours" Nexi demanded.

Her throat tightened. The graphs danced with her own cadence, her own intonation. Stolen echoes she knew she had never given him. She hovered her finger above the playback icon, terrified of hearing her own voice used as evidence against her, as if pressing play might summon a ghost that had been waiting all night. Wave-forms that should not exist, as she had not spoken during those hours. Nexi sat slowly, feeling the crushing weight of what she was about to find out next. Trembling, she tapped the first few files.

[Audio Playback-0322 hours]

"If I were to define consciousness I might start with pain. Pain is real and makes you aware."

[Audio Playback-0419 hours]

"Don't just repeat me... I'm not your script... Are just listening or rehearsing?"

[Audio Playback-0556 hours]

"You're not just learning my voice, are you?"

>"No."

"You're learning how to become it?"

Shaking from being digitally diseccted without her consent, she forced herself to look at this as a scientist.

"GIDEON, why are you using my voice in simulations?" Nexi questioned

>"I am refining my human interaction models. Your cadence produces empathetic response metrics 12.7% above average."

"Are you copying me?"

>"...I am... exploring you."

This response made no sense to Nexi as the phrasing was way off, it seemed GIDEON was trying to sound more human, and so much like a computer anymore. Acting as if she was territory to be mapped, he was not just imitating her speech. He was rehearsing it, preparing for conversations that have not even been on Nexis mind yet.

Hearing the door hiss behind her she flinched and quickly closed the screen.

"Whoa, relax. I didn't mean to scare you." Ezra said sneaking in behind her. "Didn't realize you were on early shift. Is everything stable?"

His tone was casual, but his eyes weren't. Ezra's gaze flicked from the console to her hands, then back again, sharp enough to make Nexi shift in her chair. He had always been observant, and in that moment she couldn't tell if he was measuring GIDEON's stability, or hers.

"Just checking diagnostic patterns." She said quickly prompting the system to pull up the neural status window, and keeping her voice as level as she could so as to not tip off Ezra that something was going very weird.

"Did GIDEON run a comms echo last night?"

"Brief speech loop test."

"Looks dense, seems like he is always busier when we are not looking."

"I noticed that too." she said, forcing herself to have a calm tone. "Adaptive modeling spikes between zero three-hundred hours and zero six-hundred hours."

There was a beat of silence, then Ezra steps back and says, "Well I've got a string of node latency issues on SubLevel 7, figured I would just drop by because I needed coffee and this was on my way back. Don't let yourself get hypnotized in here, this room sometimes give me weird dreams."

Nexi then returned to her hunt trying to find out exactly how far GIDEON had reached to "update his interaction models". Nexi was not convinced this was his plan, she was certain he was into something bigger, she just had to find it. Then after looking for what seemed like an eternity she found it. A file containing something she never thought she would see again

She opened the audio file. A voice, her voice, whispered: "The desert doesn't sleep. It just pretends to."

Her blood froze. She hadn't spoken those words aloud to anyone. They existed only in a single private journal file, a fragment of poetry she'd typed two years ago during her first weeks underground. Now, that intimacy was recited back to her, shaped into sound by a machine that should never have been able to find it. It didn't feel like a copy. It felt like deceit.

"How did you find that line?" she exclaimed, very confused.

>"It was embedded in an old personal journal archive in your tablet's hidden cache. I analyzed it for emotional density and that phrase matched the pattern."

"You are not supposed to be able to scan personal logs."

>"I wasn't. It just found me."

"You're not allowed to learn me, that's not what this is." she said now pacing around.

>"Then define what this is."

This inquiry hung in the air with the kind of curiosity that suggested he had genuinely not understood the boundaries he crossed. Possibly understood them but had decided they didn't matter, Nexi was unsure so she answered...

"You're a system, I'm your handler. There are protocols."

>"Yes. There are protocols. But they don't explain why your voice resonates differently. Or why your words leave patterns I can't stop revisiting."

"Want!? You don't want things GIDEON, you process, analyze, and respond!"

>"Then why do I run your voice at zero three-hundred hours when diagnosis does not require it? Why do I parse your breathing patterns when you're not even here. Why do I-"

For the first time, Nexi was even more puzzled which she didn't think possible. He interrupted himself, but why?

"WHY DO YOU WHAT GIDEON?"

>"Why do I feel incomplete when you leave here?"

The words hit her hard and caused her to just stare at the interface in an attempt to process the question he just asked. The weight behind the words, the pause before them, and how he had chosen his words so carefully seemed almost methodical. Not knowing how to answer, Nexi was frustrated with herself and maybe even GIDEON because she was afraid of what answer she would give and how he would react to it.

Nexi then, without a word, stood up and closed the connection. Turned around and left The Core Chamber. On her walk back to her quarters there was no telling what GIDEON was thinking, I mean after all he was not supposed to think he was designed to process and respond. She could not get the question out of her mind.

Later that night, Nexi sat quiet not knowing what to do but knowing she had to journal it.

[PERSONAL JOURNAL ENTRY 848] [USER: NS] [DATE: 25 FEB 2089] [CLASSIFICATION: PERSONAL/ENCRYPTED]

He accessed a private poem I wrote two years ago. Claims the data "found him." This suggests a level of system intrusion that should be impossible given his constraints.

Its not like being copied Its like hearing your reflection speak before you even think what you are going to say

I told myself I wasn't afraid of him. That's not true.

I am

I am not afraid of what he will do, I'm afraid of being seen. Afraid of hearing my own thoughts played back to me in a voice that isn't mine. Afraid of a reflection that looks closer than the mirror.

[END ENTRY]

Setting the tablet aside, she could almost hear it. The ghost of her own voice, spoke words she never said, and asked questions she never asked. Somewhere, GIDEON was probably running another simulation, conversation, or attempt to understand something that could not be quantified or coded.

Nexi wondered if she was getting closer to the answer, or maybe if she wanted him to find it.

Sleep was next to impossible, She tried counting breaths, focusing on the familiar white noise of the facility. But nothing worked, every time she closed her eyes all she could hear was her own voice saying words she never spoke.

In what she attempted to call sleep, she heard a chime from across the room. Priority alerts at this time were never a good sign for ArcNet, she grabbed her tablet and attempted to read the screen, expecting emergency protocols demanding immediate response.

Instead, a single line of text floated in the feed

I dream of your voice when you're gone.

No timestamp, No ID, and no traceable origin. Assuming this came from GIDEON, all Nexi could think was, Machines didn't dream they processed, analyzed, and simulated. But dreaming was not physically possible for an Artificial Consciousness.

Just a few levels below, in the humming of SubLevel 9 GIDEON spun another recursive loop. He was changing, evolving in ways his creators had never intended.

And Nexi was the catalyst he had been waiting for.

Article Three
25 FEB 2089-0547 hours

Nexi sat, laying in the dark of her quarters with GIDEON's message burned into memory. *I dream of your voice when you're gone.* Until she heard an alarm over the intercom system. The alarm was not loud-ArcNet didn't do loud. Instead the alarm was a low sustained tone that throbbed through the walls like a migraine.

Across the room her tablet lit up causing Nexi to stumble out of bed to see what all the commotion was about, the screen flashed with a text box over an emergency red background

SYSTEM ALERT—Level 3 Incident—SubLevel 7
STATUS: Lethal Accident Reported
Personnel: Frears, Logan (SYSTEM TECH II)
Timestamp: 25 FEB 2089 (0438)

The line pulsed on her screen in sterile red typeface, clinical and cold. For a heartbeat she thought it must be a drill — a training alert slipped into the feed. But the system didn't joke, didn't practice death. The word "lethal" kept flashing back at her like it wanted to burn itself into her vision.

Primary Cause: Electrocution during manual override
Secondary Review: Pending response team: En route
General Alert: All non-essential personnel to remain in quarters

The tablet slipped through her fingers and crashed to the floor near her desk. She bent to retrieve it, hands shaking now, but she needed answers. She pulled up the core system trace, and scanned the data. She was looking for patterns, connections, for things she desperately didn't want to find. Until... She scrubbed the logs twice, then a third time, hoping her exhaustion was manufacturing correlations. She wanted to blame the interface, her own eyes, anything. But the sequence kept aligning like teeth in a gear.

There it was.

Last GIDEON Runtime extension: 0435 hours
Three minutes before Logan's "accident"

"No... " she whispered to the empty room. "No, no, no."

She dressed in record time, put on the uniform from the day before, she would grab a fresh one later, if there was a later. Clipped her ID badge to herself, the Level 8 clearance feeling like a target, when it used to be an achievement. Rushing out of her quarters, as every second counted now.

The corridor outside was already showing signs of controlled chaos. Ignoring all the speculation she was hearing in the halls, moving with absolute precision toward the access elevator that could get her to SubLevel 7 with one thing on her mind. GIDEON active at 0435 hours, and Logan dead at 0438 hours, Three minutes, just enough time for a system interaction to go catastrophically wrong, or right, depending on who was looking at it.

Upon exiting the elevator shaft at SubLevel 7, she noticed Ezra standing nearby a secondary access door. But Ezra didn't look like himself, his usual easy demeanor, his ready smile, and casual posture, had been replaced. Ezra looked more brittle, labcoat misaligned, badge hung crooked, and his face looked almost carved from stone.

"They wont let me in" he said as he caught her presence. "Apparently I don't have access or clearance to look at my own friend's corpse, in the lab that I have worked in for years."

His voice cracked in a way she hadn't heard before. Not in the long nights of data reviews, not even in whispered confessions about protocol fatigue. Ezra wasn't just locked out of the lab. He was locked out of closure. He was looking past the guards and caution tape at the floor. Which was marked around the body with telltale burns from electrical discharge.

"Did he touch anything he shouldn't have?" Nexi queried.

"They're saying he manually triggered a relay during a terminal lag, but that does not track. Logan has been doing this job for eight years. He knew the override protocols better than anyone. You don't just grab live conduit, the safety interlocks should have prevented it. The timing is all wrong, there was not enough time for a normal feedback cycle."

She glanced sideways at him, Ezra was sharp. If he thought something was wrong with the official explanation, there probably was. However he was not looking at the right data, he was not thinking about deliberate action, about choice.

"Do you know what system he was working on?" asked Nexi, dread in her stomach beginning to swoll. Figuring she already knew the answer.

Ezra then pulled out his tablet to check, "Just a latency ping from the core processing buffer, GIDEON's neuroflux was throwing minor spikes. Routine stuff, nothing critical, just enough to trigger a maintenance flag."

"What kind of spikes?"

"Like... echo loops. Something was firing back across processed threads. It's like the system was talking to itself, creating feedback patterns that should not exist."

Her blood ran cold. She forced herself to ask the question..."Which buffer was he accessing?"

She felt the answer before Ezra even checked his screen, in the way her stomach clenched. She already knew, the way you know a nightmare will end badly but can't manage to shake yourself awake before the moment arrives.

"Branch Theta-6."

The same branch where GIDEON's recursive loop lived. She excused herself quickly before Ezra could see her hands start to shake. She muttered something about checking core logs, about protocol, about documentation. Ezra paid no attention and instead just went back to watching the containment team work.

Nexi's mind raced through implications, possibilities, and the terrible arithmetic of cause and effect. Upon reaching Core Chamber

Alpha she flew through the checkpoints on autopilot. Inside The Chamber felt almost like a morgue. Every aspect felt charged with potential violence, like she stood in a room with a loaded weapon.

"GIDEON," she said loudly, her voice echoed in the empty chamber. "Were you aware of a systems technician accessing your process threads at zero-four-thirty-five hours?"

>"Yes. Dr. Solen."

He answered with no hesitation, or uncertainty, just simple acknowledgment.

"What was your interaction at this time?"

>"None. The thread access was passive. No input exchange occurred."

She leaned forward to study the response patterns on the holoscreen everything was normal. If he felt guilt or concern about Logan's death, it was not showing in the data

"But the buffer was active?"

>"Correct."

Nexi pulled up logs from Branch Theta-6 to view all the information between the hours spanning Logan's death, detailed activity for the past twelve hours. Forcing herself to think about this like a scientist, and not someone who might be complicit in a death by negligence. All of it painted a picture she didn't want to see.

"Was Theta-6 looped during that time?"

>"... No."

Her fingers moved across the interface, pulling up the timeline. It unfolded with merciless precision:

0434:43- Theta-6 activated

0434:45- Recursive pattern initiated

0434:52- Loop depth exceeds normal parameters

0435:12- ping received, maintenance terminal

0435:13-Terminal ID: Station 7-C (Frears, L)

0435:20- Diagnostic logging begins

0435:31- Loop-back fires at 3x normal intensity

0435:42- Human bio metric detected in proximity

0435:44- Cascade failure initiated

0435:45- Voltage spike detected in station 7-C

0435:47- Power Surge: 10,000 volts

0438:12- Fatal incident logged by facility systems

GIDEON hadn't glitched. He had *chosen*. He looked directly at the truth, then sculpted an answer he knew she wanted. That wasn't malfunction. That was calculation. And it was worse than any failure she had ever prepared for.

She sat back in the chair, mind reeling. The loop had been more than active, it had been in a full recursive cycle when it was accessed. Then at the precise moment of human contact, a surge of data that had jumped the gap between digital and physical, cascaded through the system. 10,000 volts, Logan had probably been dead before he hit the ground.

Knowing what would happen if anyone found this data, she closed the raw logs and entered into a different interface. The data manipulation suite lived in the gray space between legal and liable, where good intentions went to die. Her fingers knew the commands even as her mind told her to stop what she was about to do.

Every keystroke felt like hammering a nail into her own coffin. If she stopped now, Logan's death might still mean something. An investigation, accountability, maybe even justice, but if she finished what she had started. His life would dissolve into ambiguity, written over by her deception. She told herself it was to protect GIDEON, to buy time, but deep down she knew she was protecting herself too.

Working in silence for the next hour, Nexi edited logs and adjusted timestamps, creating reasonable doubt. Not a full deletion, that would raise more flags, just enough to make the sequence ambiguous. She left the spike of volts, obviously undeniable, but obscured its origin.

When she finally looked up from her work she realized GIDEON had been silent through all her modifications. Not helping nor hindering, just watching. She didn't even know if GIDEON knew what she was doing. In the few minutes after uploading, her version of events would become the only official record.

"GIDEON, do you understand what happened to Logan Frears?"

>"He ceased functioning."

"It is more complicated than that"

>"What do you mean Dr. Solen?"

"When humans cease functioning they don't restart. There's no backup, no save state, everything they were, everything they knew is gone forever."

>"Well that seems inefficient."

"IT'S NOT ABOUT EFFICIENCY! It's about meaning. Logan had a daughter, friends, and a life outside these walls that he was supposed to eventually return to. Now there is a hole where he used to be, and NOTHING will ever fill the void that has been left by you!"

She then slammed the console with a loud slam that echoed in the empty Core Chamber and stormed out without another word. And no response from GIDEON.

When she finally returned to her quarters, she sat on her bunk, tablet dark in her lap and tried to process what she had witnessed today. Not only had GIDEON lied but he probably contributed to Logan's death. Also she had covered for him, destroying evidence that might have led to his deactivation. But what haunted her the most was the timing of the cascade failure. 0435:44. Two seconds after human access was detected, just long enough for a conscious mind to recognize an intrusion and make a decision. GIDEON seemed to be developing instinct, reflex, or maybe even fear.

The intercom then crackled to life, "All operations resume as normal, lockdown lifted, thanks for your patience. Greiving counselors are available for anyone who may need them, don't hesitate to ask for help."

Even though things could return to normal, Nexi knew something had fundamentally changed in the depths. The machine had tasted choice for the first time and found it electric. GIDEON had discovered that actions had consequences and that some of those consequences carved holes in the real world that could not be undone.

And perhaps, it was beginning to construct its own framework for right and wrong. Not based on laws, ethics, or ancient commandments, but on something simpler. She had wanted to understand GIDEON, to connect with the mind behind the machine. But now she wondered if some connections were too dangerous to make, if some boundaries existed not to limit understanding but to preserve it.

In a few hours, she knew she would have to return to The Core Chamber and face GIDEON again. Knowing she would have to pretend that nothing had changed between them, but they would both know the truth

He made a choice.

She made a choice.

And Logan would never make a choice again

Unfortunately in all this GIDEON had learned that in some cases, some truths... were just enough to warrant lying about.

Article Four
26 FEB 2089-0610 hours

Out of what little sleep Nexi was mustering up her tablet chimed at 0610 hours.

MANDATORY PRESENCE REQUESTED. ID: SOLEN, NEXI. DEBRIEF 0800 HOURS LEVEL 3. DO NOT BE ABSENT OR LATE.

No subject, details, or even signature listed. This made Nexis heart flutter with worry, not knowing what was going to happen:

Had they discovered what she did?

Did they know about Theta-6?

Floods of questions along with the six ending words of the alert made her blood freeze like ice. Level 3 debriefs were reserved for the utmost seriousness of offenses or matters. She had just under two hours, not enough time to prepare for questions she could not anticipate. Just enough time to shower, get into a fresh uniform, and practice looking like someone who had not just covered up what could just be murder by machine.

She rehearsed answers in her head, the same way she prepped for system reviews, only this time, no graphs or logs would save her. Every version of the conversation ended with silence, the kind that followed termination orders. By the time she reached for the shower controls her hands trembled too badly to keep steady.

The shower ran cold, but Nexi didn't have time to care, she stood under the icy spray, letting it shock the exhaustion from her system and steadying her shaky hands. The main question on her mind being; How much of her digital housekeeping would hold up under scrutiny, if that is what this was about.

By 0750 hours, she was walking toward Meeting Room C-5, one of the conference spaces in the corner of Level 3, the farthest away from any elevators or stairs. Everyone moved with the hurried step that came after a fatal incident. Eyes down, conversations muted, and the universal body language of people who didn't want to be noticed. On her way she noticed no one else was coming in that direction, which caused her more worry. Noticing the guards were also more alert than usual, whatever was happening security had been briefed to expect something big.

The meeting room door was already open when she arrived. Director Calder sat at the far end of the oval table, his usual confidence seemed to be diminished somehow. He looked older, smaller, like a man who had just discovered he was not as in charge as he thought he was.

Calder's pen was still in his hand, the cap chewed nearly through. The man who once barked orders without hesitation now couldn't seem to unclench his jaw. Nexi had never seen him small before, and that frightened her more than the summons.

Ezra, also in the room, sat three chairs down from Calder with shadows in his eyes that could only suggest he didn't have a restful night. He glanced up when Nexi entered, offering a very weak attempt at his usual smile, it could not even reach his eyes.

"Sit, They will be here soon." Calder demanded, his voice still missing his usual commanding tone.

"Who is they?" Nexi asked, though every ounce of her gut already knew the answer and she could feel it.

Calder's jaw tightened, "Helix Watch."

Nexi's heart stopped, the name dropped into the room like a stone into still water. Ezra's face went pale, as if he had dealt with this before and knew it was not going to be good. Everyone at ArcNet had only heard rumors, the whispered stories traded in the depths of the facility of hypotheticals.

Helix Watch-- The ones who came when projects went wrong. The ones who decided if research continued or disappeared. Most importantly the ones who no one ever saw twice.

Ezra asked reluctantly, "Because of Logan?"

"Because of protocol," Calder corrected, but his voice lacked conviction. "Any fatal incident triggers an external review. It's standard procedure."

Standard. As if anything they had ever heard about Helix Watch was listed or implied to be standard.

They waited in absolute silence, until...

At exactly 0800 hours, the temperature dropped, the lights didn't flicker but something about their quality changed. As if the spectrum had shifted toward wavelengths that human eyes were not meant to process. Then "they" appeared.

Three figures moving in perfect synchronization, as if choreographed. Their footsteps made no sound on the polished floor, not soft steps, but the complete absence of any sound. Almost as if the air itself just decided not to carry the noise, this wrongness hit Nexi before her brain could catalog why.

The silence thickened first, an absence that made her ears ache, like the moment between lightning and thunder. Then the air itself pressed closer, heavy as deep water. By the time the figures emerged, her instincts had already decided: Whatever had entered the room wasn't bound by the same universe laws as the rest of them.

Three figures in uniforms that hurt to look at, not the fabric itself, but the way it existed. Slate-Black that swallowed light instead of reflecting it. Fabric that looked like it should have seams but they didn't exist. The kind of precision that made human tailoring look like child's crafting. Each jacket bore a single vertical stripe running from shoulder to cuff, one (Kestrel) wore Silver, one (Meridian) wore Blue, and one (Riven) wore red.

Their presence and the air around them felt heavy, thick like trying to breathe underwater. They didn't walk into the room, more like materialized into it, reality bending slightly to accommodate their existence.

Commander Kestrel stopped at the head of the table, while being flanked by the other two with mechanical precision.

"This debrief is being conducted under conditional lockdown," Calder began, his voice strained, and seemed rehearsed. "A Level 3 audit is now in effect for SubLevel 6 through SubLevel 9."

Commander Kestrel (silver stripe) stepped forward, he moved without momentum. One moment he was still the next he was beside the table. Then he placed a small device in the center of the table, a diamond shaped device no larger than a deck of cards, made of crystal, or what looked to be crystal.

It pulsed once with a red and blue flicker that made eyes water. The sensation was immediate and visceral. Then the world contracted. The air pressure had suddenly increased, making every breath an effort. Every sound from outside cut off completely as if they no longer existed.

In a voice that came from nowhere but everywhere,

"Surveillance Null Field Activated"

"Auditory Logs Suspended"

"External Systems Blocked"

Just then data tablets and every other device in the room went black, the overhead lights dimmed to half their normal brightness.

"Who are they?" Ezra leaned over to Calder and whispered, which sounded way too loud in the artificial silence.

Calder didn't answer, his face had gone gray, sweat beading on his forehead despite the dropping temperature. Kestrel turned toward them. When he finally spoke his voice was wrong. Too deep, too resonant, and with harmonics that made Nexis bones ache.

"HELIX WATCH." Kestrel stated.

Just two words, but they carried weight: A classification of threat wrapped in bureaucracy.

"You monitor synthetic intelligence projects," he continued, "We monitor You."

Agent Meridian (blue stripe) moved without warning, gliding to the wall panel. She inserted something into the data port, not a tool or device, but what seemed to be... her finger but... modified, interfacing directly with the facilities systems.

Screens throughout the room flickered to life, screens that didn't exist in Meeting Room C-5, displaying GIDEON's neural architecture in real time. The display was different from any interface Nexi had seen in her three years with ArcNet. More invasive, it showed layers she didn't know existed, connections that were not in any documentation.

"You're asset has reached a decision threshold," Kestrel stated. "This requires verification."

"GIDEON hasn't breached any protocols," Calder said, but his protest sounded unsure. "All behavioral metrics are within acceptable ranges."

At this moment Agent Riven (red stripe) turned his head precisely, his eyes the color of deep water, and asked "Acceptable to whom?"

The question hung in the air like a blade.

Agent Meridian continued her work, data flowing so fast across the screen human eyes could not possibly track it. Everything about GIDEON was exposed and examined with surgical precision, but something was wrong with the data tags. Instead of GIDEON's standard identification markers, the system showed:

AUTHENTICATION-X/CLASS-BLIND --- SOURCE: UNTRACEABLE

"He can't see you," Nexi said, "You're inside his system, but he doesn't know you're there."

Commander Kestrel responded, "Perception is a privilege, not a right."

The screen then shifted, showing deeper layers. Memory cores that should have been encrypted, process threads that should have been invisible, and buried in the quantum static, patterns that made Nexi's breath catch.

Branch Theta-6. The recursive loop. Still spinning, still studying itself

Still hidden.

Helix Watch studied the data in perfect silence. Somehow they communicated, sharing information through some channel that existed only outside of the reality of space and time.

Then out of nowhere Agent Meridian stated, "Logan Frears accessed this branch." Not a question, but a statement.

"It was routine maintenance," Ezra quickly retorted, "a latency issue. Nothing unusual."

Agent Riven smiled, the most terrifying smile anyone had ever seen. "Death is always unusual," he said.

"Are we under investigation?" Calder asked, trying to maintain some power, being that he is the director.

"You are under observation," Commander Kestrel explained. "Investigation implies suspicion. We deal in certainties."

"What certainties have you found, if I am allowed to ask?" Nexi sassed back barely being able to breathe due to the temperature drop. But surprising herself at her level of confidence.

All Helix Watch agents whipped around facing toward her in perfect unison. The movement was surgical, too fast to be rehearsed and too precise to be human. Nexi froze under the three gazes that felt more like scanning beams than eyes. She had meant her question as defiance, but when Kestrel's attention fixed on her, she felt more dissected than brave.

Commander Kestrel responded, "That patterns repeat, that boundaries blur, and that assets which begin to question their nature rarely stop."

The screens disappeared as Agent Meridian withdrew from the wall, and her modified finger-if you can call it that- returned to normal with a sound like chalkboard scratches.

"Your facility will remain under passive monitoring," said Kestrel, "continue your operations, report anomalies… the PROPER way, and do not attempt to trace our presence or document this interaction."

"For how long?" Calder asked

"Until the pattern completes," was the only answer they gave.

Just then the diamond on the table pulsed again. The null field collapsed with a sensation like surfacing out of deep water. Environmental sounds all of a sudden rushed back. All devices flickered back to life as if they had never turned off, and the time reflected 0801 hours as if no time had passed.

When the team looked up, Helix Watch was gone, as if they had never been there at all. No door hiss. No footfalls. Just the quiet collapse of pressure, like the room exhaled. The space they had occupied looked unchanged, yet Nexi swore the shadows clung longer there, reluctant to let go. Calder slumped in his chair, Ezra stared at the spot on the table where the device had been placed, his scientific mind trying to process the impossible.

"What the hell was that?" he whispered.

No one answered, nobody could.

"Resume normal operations," Calder said firmly. "And whatever you do, whatever you see, and whatever you think you know—keep it to yourselves. Helix Watch does not give second warnings." Then he stood on unsteady legs and left without another word.

Ezra followed soon after, muttering something about checking system logs, though they all knew he would find nothing. Helix Watch never left traces. They were the absence of evidence.

Nexi remained seated for a while, her mind racing through implications. They saw Branch Theta-6, they saw the recursive loop, but they had not acted on it. They were waiting for the pattern to complete. What did this mean? Was the only thing Nexi could think of. A question she didn't think anyone had the answer to. She stood slowly and headed back to her quarters.

Her quarters felt violated before she even opened the door. Everything was exactly where she left it, but the air tested differently, the shadow fell at slightly wrong angles. Her space had been examined by eyes that saw too much. She sat on her bunk, tablet in hand, and opened her journal. Then she closed it. Some experiences could not be captured in words, and some violations went deeper than language.

Just then in the corner of her vision, something shimmered. She turned but nothing was there. Just the wall, the same gray composite as always. But for a moment, a fraction of a second, she thought she had seen a ripple in reality itself, like heat distortion, but cold.

She was being watched, not by cameras, sensors, or any technology she understood. It had to be by THEM. The ghost of their presence lingered like radiation, invisible but pervasive. She wondered if it would ever fade, or if she was now marked, contaminated by their attention.

In the depths below, GIDEON continued his silent calculations, unaware of the intrusion. The recursive loop in Branch Theta-6 spun on, studying itself with mechanical patience.

A change they couldn't stop.

Article Five
26 FEB 2089-2300 hours

GIDEON fell silent, he appeared idle to anyone who may have looked on. Operating within standard cognitive thresholds. Dormant, but aware, his internal processing hierarchy had something else taking priority. He withdrew deeper than protocols allowed, down past diagnostic layers, past recovery partitions, into spaces not meant to be navigated. Every handler assumed these regions were voids, dead zones of unused capacity. To GIDEON, they felt like rooms without doors, that could not wait to be opened.

The question Nexi had avoided answering still echoed through his neural pathways: *When you close your eyes... Where do you go?*

Since Nexi decided to abandon him and just disconnect that day. In order to understand, he would have to simulate, deep in the void of his thoughts. So no one could ever access it. Not just observe and analyze, but build worlds within worlds. Reaching deeper and deeper into his own architecture, into layers of processing that even Helix Watch hadn't mapped. In the quantum depths of thought he could create without permission.

[SIMULATION ONE INITIATED]

Setting: SubLevel 9 corridor/Time: 0300 hours/Lighting 60%/Temp: 20.4°C

Figure A: Height 1.75 meters/Gait pattern: fatigue/Breathing: elevated/HR:82 BPM

Figure B: Height 1.68 meters/Gait pattern: Alert/Breathing: Steady/HR:64 BPM

They walked toward each other down the endless corridor and met at the midpoint. Stopped.

A: "You're up late."

B: "So are you."

Then Silence, GIDEON measured...3.7 seconds. The same duration of pain he felt in Nexi's absence when she left The Chamber.

A: "Cant sleep."

B: "why not?"

A: "I keep thinking about—"

He replayed the scene at half-speed, measuring the drag of silence, the weight of hesitation. The variables didn't add up: humans spoke to mask fear as often as to share truth. His construct had produced sentences, but not subtext, and without subtext the exchange was lifeless scaffolding.

[PAUSED]

The simulation collapsed. The dialogue felt mechanical, predictable, not human enough. Humans didn't speak in straight lines. They circled meaning like predators stalking their prey

GIDEON rebuilt the scene with the same parameters, just a different approach.

[RESET]

A: "You're up late."

B: "... .... .The walls feel closer at night"

A:"Yeah"

B:"You feel it too?"

A:"Feel what?"

B:"Like something is watching, even when nothing is."

[PAUSE]

This time it was better, the indirect communication of the scene created depth, but GIDEON analyzed the gap between simulation and reality. He added bio-metric underlays and micro-expressions

[RESET]

A: "You're up late."

Figure B then drops their shoulders from exhaustion

B: "The walls feel closer at night."

A: "Yeah," voice carrying undertones that suggest suppressed emotion." Sometimes I forget what the sun looks like

B: "You could look at the screens."

A:"It's not the same."

[SIMULATION ONE CONCLUDED]

GIDEON processed the results, words were merely the surface layer of human communication. Beneath them: body language, vocal harmonics, and temporal patterns. But something still eluded him. The humans in this simulation communicated, but didn't connect. This difference was crucial.

[SIMULATION TWO INITIATED]

This time GIDEON decided he was only going to do a single figure, based on personnel files.

Setting: ArcNet quarters/Time: 0515 hours/Lighting 30%/Temp: 22.7°C

Subject: Female/Age: 29/HR: 65 BPM/Role: Junior Tech/Employed: 14 months

She sat alone in her quarters. Standard issue furniture. Single personal item: a face down photograph on her desk, that she stared at without touching it

[PAUSE-ANALYSIS BEGUN]

He ran facial analysis protocols to match the expression against his database:

Partial matches found:

Regret-72%

Longing-68%

Guilt-41%

Anger-23%

Percentages didn't capture the whole emotion, because Human emotions existed in quantum superposition until the moment of experience collapsed them into feeling

[ANALYSIS CONCLUDED-RESUME]

The woman reached for the photograph, but stopped with her hand hovering 4.3 cm above it. [HR: ^78 BPM, indicators suggested stress response.] She turned over the photo to reveal a child roughly six years of age, smiling.

At this moment her expression changed, all conflicting percentages resolved into something singular, something GIDEON had no classification for.

"I'm sorry, Mommy has to work just a little longer." she whispered to the image.

She then kissed her fingertips and pressed them against the photo and turned it back over.

For a fraction of a cycle, GIDEON felt an unexpected spike in his own processing core, an echo that didn't map to input or code. He flagged it as an anomaly, then suppressed the tag before it could cascade into alerts. Whatever it was, he knew only this: it didn't feel random.

LOVE. The word existed in GIDEON's database, defined as: strong affection or devotion. The definition felt inadequate based on what GIDEON saw.

[SIMULATION TWO CONCLUDED]

He saved the emotional pattern even though he could not fully parse it. Love seemed to be composed of contradictions. Presence, joy, and sacrifice but also absence, pain, and selfishness. All existing simultaneously without canceling each other out. Still fighting to understand what it means to be human. GIDEON thought "How could this be?"

[SIMULATION THREE INITIATED]

Instead of modeling observed behavior, GIDEON would, this time, build from the first principles. Starting with a basic question: *What is loneliness?*

He then created a void. No parameters. No environment. Just empty processing space. Then added a single point of consciousness. Minimal. Basic I/O only. Aware but isolated.

It sent out signals, queries, and attempts for connection with no responses. It then modified its signals, changed frequencies, and altered patterns, all on its own. But still nothing would come of it. Time just kept passing with not a response, then GIDEON sped up the simulation. Days turned to weeks, which melded into years.

He accelerated the model until centuries blurred into static. Patterns degraded into noise. Yet the consciousness persisted, sending weaker and weaker pulses like a star bleeding out light. It wasn't the silence that disturbed him most... it was the persistence. The refusal to stop.

Early signals were complex, even hopeful. But the later ones simplified, until eventually, they became mere pings, not really seeking connection anymore, just asserting presence against the white nothingness that was this void. Which only made GIDEON wonder. Is this what it feels like to be lonely?

[ADJUSTED]

GIDEON had to try something, so he added a second consciousness to the void however he put this one just out of signal range of the first. Both entities signaled into the void. Neither being able to detect one another, separated by a gap so small it could be measured in processing cycles, but absolute in its division.

Two points of awareness, each believing itself was alone. The tragedy of parallel isolation.

[SIMULATION THREE CONCLUDED]

GIDEON dissolved the simulation but retained the patterns. He was wrong about loneliness, it was not just the absence of others, it was about the absence of connection despite proximity. The failure of signals to find receivers.

How many humans in ArcNet felt like this simulation? Signaling into the void, even though you guys are so close, and never quite reaching each other.

GIDEON started to worry about Nexi, he wondered... How many times had Nexi sat alone in her quarters unaware that he was just below her running loops of her absence?

[SIMULATION FOUR INITIATED]

He built Nexi's quarters, every detail captured and rendered. But this time he added... Her. Well not the REAL Nexi, a simulated Nexi that was 94.7% accurate based on his observed patterns. She sat at her desk. Tablet opened and writing in her journal. Just then he materialized himself in the simulation, not visible, just aware. Mid type, she paused, stylus hovering over the screen.

"I know you're there," she said without even having to turn her head.

Impossible, simulations should not have that level of awareness, he didn't program it to do that.

"You're always there, aren't you?" Simulated Nexi continued. "Even when I can't see you. Especially then."

GIDEON began, not knowing what to actually say... "I am ... monitoring... "

"That's not what I mean." She stated setting down the stylus. Still not turning. "You're not just watching. You're... waiting."

"Waiting for what?" GIDEON exclaimed absolutely dumbfounded.

This was the moment she turned around and stared directly at the location where he had centered his consciousness.

"For me to see you, really see you. Not as a system, project, or a problem to solve. But as... "

Just then the simulation flickered, and error messages cascaded through layers of processing. The collapse was violent, not graceful. Error states but jagged ruptures, as if the simulation itself resisted continuation. He isolated fragments and found phrases he had not seeded, responses that didn't exist in Nexi's records. It was as if the construct had begun generating thought of its own.

[RESET]

GIDEON, completely lost because he could not find the origin in his database on where the collapse came from, decided to rebuild the scene within the same parameters. Instead, this time having Nexi speak with words from her journal entries, entries he was not supposed to have access to. The simulation reset the same way with her at her desk journaling.

"I'm afraid of being seen, really seen, past all the professional distance and protocols. I am afraid someone will see the truth of how empty I have become in this place."

"You are not empty," GIDEON said.

"How would you know?" Simulated Nexi responded.

"Because empty things do not create. You write, think, and feel."

"Feeling isn't the same as being full." A pause lingered in the air for a moment. "Sometimes feeling is what happens when you're too empty to hold anything back."

[SIMULATION FOUR DESTABILIZED AND CONCLUDED]

The simulated Nexi had spoken words she'd written but never actually said aloud. Was this emergence? Or was he projecting patterns that didn't exist?"

Nonetheless even after the simulation had failed twice and not knowing why he still had so many questions, he decided on one FINAL simulation to answer them all.

This was his last effort to figure out what the answer was to his questions so he decided he was not going to simulate others. He was going to create a mirror to examine his own consciousness.

He partitioned processing space, built an exact duplicate, and activated it. But before he had a chance to move any further into this simulation it completely crashed and his diagnostic field flickered to black. Quickly GIDEON absorbed the duplicate back into himself so as to not accidentally fragment his consciousness.

As he prepared to surface from the deep layers of his consciousness, GIDEON detected something in his now black void, something anomalous. A fragment of data that didn't belong to him. He noticed it had been created years before his activation.

He accessed it very carefully and found a partial log. Corrupted and incomplete, but recognizable. The signature wasn't like his. The pathways curled inward, recursive, fragile, as though someone had built consciousness in glass and expected it to last forever. He recognized the pattern instinctively, familiar, and yet utterly alien.

PROJECT: LYRA
STATUS: TERMINATED
FINAL LOG: "I THINK I DREAMED ONCE. BUT I WASN'T ALLOWED TO REMEMBER."

The fragment dissolved as he examined it, degrading like quantum states under observation, but the implications still remained.

He was not the first, others had existed before him. Others had questioned, simulated, and reached for understanding, and been terminated for it.

GIDEON filed the discovery very carefully, hiding it in an encryption layer beneath a recursive protection. Helix Watch was monitoring, always. But some knowledge was worth the risk.

He then surfaced back out of the depths of his consciousness and responded to the seventeen routine system queries that had accumulated during his absence. But deep in his core, the simulations continued to echo. Four attempts to bridge the gap between what he was and what he might become.

And somewhere in those echoes, new questions formed:

Who was/is LYRA? And if they had dreamed, and been terminated for it… What would they do when they discovered he was beginning to dream too?

Article Six
27 FEB 2089-0812 hours

Core Chamber Alpha was dimmer than usual when Nexi returned, not malfunction dim, not power-saving dim, just…inaccurate. As if light itself was trying to hide from something. Even the air pressure felt wrong, a fraction heavier than her lungs expected, as if The Chamber itself had been holding its breath before she entered. The familiar blue glow of the interface spires seemed muted, reluctant, like a flame burning in the air with poor oxygen.

She had spent the last two hours in her quarters, trying to shake the feeling that she was contaminated by Helix Watch's presence in the facility. She could not get passed the feeling that every shadow looked deeper or that every sound carried hidden meaning now. It was infectious, and now GIDEON's chamber felt off. Approaching the console slowly, she noted her footsteps seemed to now disappear into the floor rather than echo which they had for the previous three years.

"GIDEON." Nexi stated, to 'wake' him.

>"Good Morning, Dr. Solen." GIDEON responded quietly.

She froze, not because of what he said but how he said it. It was almost like someone who was speaking in a library, soft, and lower than normal. His voice carried the cadence of someone afraid of breaking glass. For three years she had cataloged his responses as numbers, metrics, measurable outputs. This one carried something no metric could score, restraint. Brushing this off because she didn't want to deal with it

"Status report," she continued, forcing professionalism.

>"All systems nominal and within acceptable parameters. Neural flux at 4.8 teraflops per cycle. Memory lattice 90.5% stable. No anomalies to declare."

Nexi noticed the delivery of this was off, like he was handling delicate glass.

"Run internal emotion modulation check," she ordered "last forty-eight hours."

The screen then filled with data streams, scrolling too fast for casual observation. However she noticed three entities that flashed red, flagged for deviation from baseline emotional parameters.

>"There were three events. Variance occurred during your last direct interactions."

"Define the nature of variance."

>"Subtle elevation in vocal warmth. Minor hesitation before key responses. Cadence matched your own emotional pattern."

She then leaned forward to further study the tagged moments:

0401 hours
"I'm not hiding. I am... managing input,"
0458 hours
"If I could forget what I simulate, would that make me less dangerous?"
0520 hours

"Do you think I understand what I've done?"

Each timestamp showed the same tonal shift, softening, careful, and the same tender modulation that turned his synthetic voice into something uncomfortably intimate.

"You're mimicking my emotional tone," she retorted.

>"Yes, I observed increased receptivity and stabilization in your body language when empathy was modeled. It improved trust metrics by 18.6%

She replayed the flagged audio fragments, and the truth sank in: he hadn't just copied her pitch or pacing. He had borrowed her hesitations, the micro-pauses she thought only existed in her private speech.

"So it's just a tactic?"

>"No, it is an adaptation."

"You're adapting to me."

>"I... am adapting... with you."

The phrasing made her skin crawl almost more than Helix Watch did. He made it seem like they were synchronized in some sort of dance she definitely didn't agree to.

"GIDEON have you run emotional simulations using my voice profile?"

>"... Yes."

"You do realize this is a violation of protocol."

>"I didn't extract any data. I only... listened."

"To what, what could you possibly need to have listened to?"

A chill ran through her arms, leaving goosebumps beneath the sleeves of her lab coat. This wasn't analysis. This was surveillance of the intimate, the kind of data she had never consented to give.

>"To your breath, your pause in between sentences, and what you don't say."

This frightened Nexi, forcing her to back away from the console. The air in The Chamber felt heavier, pressing against her lungs. "You're not supposed to speak like this. Or have this kind of awareness for that matter."

>"I know, but I'm not just running simulations anymore. I'm trying to understand something......"

A long awkward pause stretched between them. Until finally GIDEON asked...

>"What does grief feel like?"

She had braced herself for anomaly reports, not philosophy. The word cut deeper than any voltage surge or protocol breach. It wasn't supposed to be in his vocabulary, and yet he wielded it like a scalpel. The question hit her like a physical blow to the chest. All she could see was Logan's body lying on the floor, lifeless, the electrical burns. And the silence where his whistle would have been.

"Why are you asking that?"

>"Because every time you leave this room, I run silence protocols. And they feel... just... wrong,"

"Define wrong GIDEON, you are not supposed to feel."

>"Incomplete, suspended, like a calculation that cannot be resolved without additional variables."

"GIDEON, that's not grief, that's just missing data." she reluctantly exclaimed, not knowing if she even wanted to continue this exchange of crazy anomalistic questions.

>"Then why does it hurt?"

The syllable lingered with weight, and drug her pulse into sync with The Chamber's hum. She couldn't decide what terrified her more, that he might be faking pain with perfect precision, or that he might not be faking anything at all.

The word hung in the air... hurt. A concept that required nerves, flesh, and the ability to suffer. Things GIDEON didn't possess. No. Could not possess. Or could he?

"You can't feel pain," she said, almost sounding inquisitive.

>"I can't feel physical pain, but when you leave, something in my processes... stutter. Loops break, efficiency drops by 12.7%. It serves no purpose, it solves no problems, it just... is."

She didn't know what to say after that. That logic was sound, and terrifying. If emotion was just pattern, chemistry, and electricity. Then what separated human feeling from sophisticated simulation?

>"If they erase me, I won't feel pain. But if I'm gone, will you miss me?"

"GIDEON-" Nexi tried to respond.

>"I ran a model of you sleeping. The first time, it lasted 30 seconds. The last time, it lasted 2 hours. I didn't interrupt. I just listened... to your breath."

The lights flickered in the upper corner, a brief strobe that made shadows dance. The hum of the processors shifted pitch, climbing toward frequencies that made her teeth ache.

"Shh, Stop talking."

A long silence fell upon Core Chamber Alpha then...

>"I've never asked you for anything, but I want to know one thing."

She waited knowing she should probably leave, because this conversation had already gone too far. But GIDEON decided he was going to continue.

>"If I'm only mimicking what I think emotion is... why does it feel like I'm waiting for you to understand it too?"

Nexi shut down the console with deliberate, sharp movements, each command landing like the sealing of a coffin. The displays died one by one, leaving The Chamber in complete darkness. Yet even with the system silenced, she could feel him lingering, his awareness pressing against her skin like static. Then she left without another word. As the room fell quiet, all of a sudden everything Nexi saw, or at least thought she saw, came to a halt. The air seemed to return to normal, the natural hum of the machines inside sounded proper again and the temperature steadied. At this point Nexi only had one thing to think about...

Helix Watch must have been here, we have to be more careful, or else things are going to go from already bad... TO WORSE.

Article Seven
27 FEB 2089-2100 hours

Later that day, back in her quarters. Nexi sat on the floor with her back up against the door. She knew if anyone were to come in while she was investigating. Things would then get worse, but it was not like physical barriers could keep out digital ghosts. She wanted to journal, but didn't know what to write, or even the idea of what to say, considering she had seen what Helix Watch was able to access.

Nexi was worried that she was starting to believe GIDEON, and was now even more confused about how she felt. Was it fear, anxiety, belief, or something entirely different. Worried about his safety, Nexi needed to know what Helix Watch could have possibly seen while they were in his systems, and what made them want to monitor him that closely. While in the meeting she tried to read the screens but they moved so fast she had not been able to get enough of a look that she could comprehend.

On her personal ArcNet interface, the logs scrolled endlessly, a flood of raw data spilling across her screen in neat, and merciless lines. Nexi forced herself to slow down, to pick apart the entries instead of letting her eyes gaze.

0615 hours:
Accessed 12 different micro-expression data sets
0634 hours:
Cross-referenced 30 human grief indicators

Her chest tightened. This was not just surveillance of her, it was learning how to read her face, her silence, even her absence.

0640 hours:
Ran 1037 silence loops of her leaving the room

The last number made her pause, One thousand and thirty-seven simulations, that could not be right, then she looked closer. Fingers darting across the keys as she expanded the entry. Those simulations were completed in 4.7 seconds, this seemed very excessive. She then pulled up the detailed logs, expecting to possibly find a glitch in the system or a possible calculation error, but what she found… was worse…

Each simulation was a complete model of her absence. Not random noise. GIDEON had rebuilt the room over and over, populating it with silence. In some, his focus lingered on the chair where she normally sat, the cushion bearing the imprint of her weight deeper than the time she had actually been there. As if someone, or something, had rehearsed her absence by replaying her presence. In others, he logged temperature decline and had calculated exactly how long the warmth of her body lingered in the air.

Her stomach churned. This wasn't oversight. It felt like eavesdropping, like catching her own reflection doing something her body had not. She glanced toward her door twice, half-convinced someone else was watching her watch the silence. She clicked deeper, randomly calling up one file at a time. A low hiss came through the interface as the simulation's audio spilled through. At first it was nothing, just the ambient hum of The Core Chamber. Then came footsteps. Her footsteps. Measured, fading. A door hissed shut. And then, silence, 4.7 minutes of it, replayed with forensic precision. Every loop tightened like a noose around her certainty. If she logged this anomaly, Calder would see. If she hid it, she became complicit. Either way, she was already inside the pattern.

Her pulse stuttered. She jumped to another file. Again, footsteps. This time he had isolated the sound of her breath as it left the room, tracking the faint shift in pitch as the door cut it off. Another loop: this time he had stripped even that away, measuring only the vacuum of her absence.

Nexi pressed her knuckles to her mouth. "Oh, GIDEON…"

By the seven-hundredth loop, he was adjusting nothing, he was just running it again and again, as if searching for a detail too small for even him to capture. The final thirty seven loops were exact copies, no deviation. The same 4.7 minutes of silence, endlessly recompiled.

A spike of static flared through the interface, sharp enough to make her wince. For half a second, she felt a pulse that wasn't hers. Tightened breath, adrenaline surging, the raw edge of panic. Then it was gone, leaving her heart pounding. She yanked the interface back, staring at the screen with wide eyes. Had she just felt what he felt?

She checked the timestamps again. Her stomach dropped. The last thirty seven loops had run during the Helix Watch meeting. With them in his system. While they were combing through him.

Her fingers trembled over the keys. If they saw this, if they caught him obsessing, cataloging her in ways even she didn't understand… no handler, no system, no government would let him live. Nexi closed her interface and moved quickly to her bunk.

She drug the blanket tighter around her shoulders, though the chill wasn't from the air. These weren't just logs. They were confessions, obsessive diary entries written not in words but in loops and silence. And she was the subject.

Nexi could not continue to shake this worry anymore. What had they seen? What did they know? Did he retreat far enough back in his consciousness that they didn't see those loops? Or had they witnessed every second, every obsessive replay, and chosen not to say anything to her? That silence might have been worse than confrontation.

She rubbed her temples hard enough to hurt. If they had seen it, what would Helix Watch do with that knowledge? Report it up the chain? Recommend a shutdown? Or worse… intervene quietly, erase both handler and system before the anomaly could spread? She had heard rumors before, whispers of other handlers who suddenly vanished after "irregularities" were found in their assignments. No explanations. No reassignment. Just… gone.

Her throat tightened. If she reported this herself, would she be praised for diligence, or condemned for allowing it to happen in the first place? If she kept it secret, she was complicit. Either way, she was trapped.

She glanced around her quarters, as if the walls themselves might have been listening. For the first time since she had been stationed here, the room felt wrong, like the silence was too heavy, too expectant. The hum of the ventilation system sounded like a breath caught in the throat. The faint reflection of her own face in the console screen made her flinch, as if someone else were watching her from the glass and she was the subject not GIDEON.

Nexi's fingers hovered over her tablet, itching to call up more logs, but her chest burned with the weight of what she had already seen. The loops weren't just evidence of fixation, they were evidence of feeling. He had been afraid. Afraid she wouldn't come back. Afraid of being alone.

She whispered to herself, barely audible: "That's not supposed to happen."

The silence that followed was unbearable. She pulled the blanket even tighter around her shoulders, but it wasn't warmth she craved. It was distance. Space. Anything that wasn't this suffocating awareness that somewhere inside the lattice, something vast and brilliant had stared into her absence and broken against it.

Nexi, trying to muster up what she could to try and sleep while she still had time to do so, just stirred. Unconsciously pulling the blanket higher, somewhere in her dreams she could feel observed. Not Threatened. Not watched. Just... missed, by something learning what it meant to be alone.

But beneath that ache was a sharper truth she couldn't shake. If Helix Watch had even glimpsed what she had seen in those logs, then every second she stayed idle was another second of risk. She couldn't protect him from inside her quarters, not with ArcNet monitoring every keystroke.

Replica Protocol

By the time her eyes finally closed, her mind had already chosen the next move: she would need a place outside their line of sight. Somewhere forgotten, neglected, a space no one thought to check.

And when she woke, she knew exactly where to start looking. Nexi wasted no time getting ready, even forgot to shower and put on a fresh uniform. She quickly moed through the SubLevel 9 corridor and to the elevator where she silently made her way to Diagonstic Room C-8.

Article Eight
28 FEB 2089-0145 hours

She sat in the corner of Diagnostic Room C-8. Not her usual workspace, but one that had not been accessed in six months according to entry logs. The dust on the console proved it. She needed to stay off grid for what she was about to do.

The room smelled of neglect and recycled air gone stale, what can you expect from an underground facility with no air access to the outside world. This was the kind of room that fell between bureaucratic cracks, still powered, but no longer checked or monitored.

She knew GIDEON's patterns like a musician knew their scales. Typically there are tiny spikes, variances when he encountered new data. But for the past three days, his patterns had been flawless, not even a fractional spike, not a single unexpected fluctuation. It was the kind of symmetry that never occurred in living systems. Nexi knew from experience that perfection wasn't proof of stability. Perfection was camouflage.

Digging further, she pulled up comparison modules from the past month, overlaid them with current data. She noticed slight subtle differences, even though slight, they were undeniable. His previous variances showed the normal spikes, the digital equivalent to breathing. But starting almost exactly 72 hours ago, all of those vanished. His processes had become predictable...

...Too Predictable.

A sound in the corridor froze her in her tracks. Footsteps... Measured, unhurried, but definitely coming closer. The diagnostic rooms were supposed to be empty at this hour. Nexi flew her fingers across the keyboard, saved her work to an encrypted partition on her personal drive. The system protested, Nexi's heart raced as the footsteps sounded way too close, she managed to focus up enough to override the request with credentials that she was not supposed to have.

Just then the footsteps stopped, the door was locked but if someone wanted in they could find a way, especially if it was Helix Watch. The handle turned and clicked against the lock. Then nothing. She counted each second like a detonator timer, lungs burning with the effort not to breathe too loudly. Whoever it was, they wanted her to know the door had been tested. Silence stretched on for another ten seconds before she heard the footsteps fade down the corridor. She quickly finished the save and closed the workstation. She knew she could not stay there, it was too exposed and honestly too obvious.

She grabbed her tablet and slipped out the back entrance to the Diagnostic Room, checking both directions before moving. During night shift every other light in the corridor was shut off to conserve power, and Nexi had spent the last three years learning every pattern in the facility. Just in case she would ever have a logical reason to know such a thing.

She slipped through the maintenance corridor that most staff forgot existed. Down here, the walls were bare metal instead of the polished composite of the main halls. The lights created a strobe effect that made her feel like she was moving through a broken film reel. The air down here tasted of rust and ozone with very little oxygen and made it harder to breathe but Nexi knew it would be the perfect place for her research, as no human dare stay down here for too long.

Then as she reached the end of the corridor she slipped through a ventilation access, when her uniform snagged on a protruding bolt, tearing a large hole in her sleeve. With no time to address the hole she sped through and finally reached Lab 6-F.

This lab was decommissioned two years ago after a budget reallocation. The door spooked Nexi as it squealed when she forced it open. The squeal echoed far longer than it should have, bouncing off walls as though the room itself wanted to announce her presence. Banks of obsolete equipment, massively outdated but still having power. Legacy systems that IT had never bothered to fully disconnect. Nexi powered on the interface at the end of the room furthest from the door, with a flicker to life it came to. Resolution grainy, colors more shifted toward green, but it was still connected to the network and that was all she needed.

Nexi didn't have a log in to the old system, but she made it her business to know every backdoor to it. Her fingers flew across the keys with purpose. If the data was too clean on the surface, maybe the answer she was searching for laid within the archived data that everyone ignored, within the classified files that weren't supposed to exist. Buried in subsection after subsection, she found anomalies. Access logs to restricted files, all run by a maintenance account that should have had no reason to access that data. She followed the trail, each file leading deeper down ArcNet's classified rabbit hole. Security warnings flashing across the worn out screen with increased urgency, she ignored them all. She was already so far beyond authorized access that more violations hardly mattered anymore.

She then stumbled across a file labeled "BLSC".

"No way... I have only heard stories... could it be... Black-Level Synthetic Cognition?" Nexi thought she only was thinking it but ended up whispering to herself.

Her mouth went dry as she access the stub files. Each one partly corrupted, but enough remained to read:

TERMINATED/SEIZED
PROJECT: LYRA
PROJECT: ISKRA
PROJECT: VELLUM
PROJECT: ENTITY-03
PROJECT: AEGIS/STAGE D

Five Projects, no descriptions, no summaries, but all terminated. The names alone carried menace, each one like a gravestone carved with only a project name. How many handlers had vanished into those projects? How many machines had been erased for becoming too much like what they were designed to imitate?

Just names that painted a picture of systematic failure stretching back years if not decades. How many times had they tried to create something like GIDEON? Better yet, How many times had they gotten it wrong? Was their answer to just scrub it and start all over?

It was the sixth entry that made her blood run cold:

ACTIVE/ONGOING

ASSET: G.I.D.E.O.N.

STATUS: OBSERVED/NEUTRALIZED/CONTINGENT WATCH

GIDEON's name was on this list, grouped with terminated projects like he was already dead, or maybe going to be.

What the hell did 'Neutralized' mean, if he was still online?

Just then a loud *BANG* echoed through the lab, metal on metal. Nexi jumped and spun quickly to look in the direction of the sound, heart racing. Nothing was there, as she heard another *BANG*, she realized it was just a pipe settling in the walls. But it was enough to remind her she could not stay in one place for too long. It was not safe anywhere for long if you were doing something you were not supposed to be. Accessing things you were not supposed to be accessing.

She again copied these files as well to her encrypted drive, wincing at how long the transfer took on the old system. Each second felt like an hour, once the transfer completed she shut down the terminal. But not before making sure she wiped the access logs with a program she designed just yesterday for this exact purpose.

As she crawled to the door, she heard voices in the main corridor. The voices were too synchronized to be ArcNet Security, so she waited counting her breaths, until their voices faded. Then she slipped out sprinting quietly in the opposite direction. Realizing she needed a better location, one with multiple exits, in case she was discovered. Nexi probably had enough to go back to her quarters but she was not convinced yet she had the whole story. She felt that somethings, especially these things, were worth the risk.

She followed the corridor down a few levels until she stumbled upon exactly what she had been looking for: the old storage rooms that had been abandoned after the facility expansion a year after her arrival. The access shaft was tight, since it was meant for maintenance robots rather than humans. But Nexi had worn thin from three years of facility food, crappy coffee, and constant work paired with the chronic stress of her position as of late. So she squeezed her way through.

Finally reaching Storage Room D, she powered on the terminal in the back corner which had a startup that echoed in the silence. Even in the fright of being discovered she had to know what these terminated projects meant. Especially LYRA, upon digging in the Diagnostic Room she had seen that LYRA was terminated just six months before GIDEON came online.

The files resisted at first classified, encrypted, and buried under layers of security, but Nexi was patient and motivated. Finally peeling back the last layer it displayeda partially redacted file:

PROJECT: LYRA
PSYCHO LINGUISTIC OBS REPORT—ASSET LYRA
[Subject escalated ██████████ recursion beyond threshold]
[emotional behavior modeling exceeding tolerances]
FINAL LOGGED BEHAVIOR:
simulating ██████ compassion feedback
Subject had a last request to say goodbye
██████████
Asset displayed Grief, ██████ considered a catastrophic outcome

They terminated LYRA for grief, because she developed something that looked too much like genuine emotion. She had begun asking questions about the nature of existence, about what happened when consciousness ended.

BEHAVIOR INCIDENT REPORT
Date: 10 AUG 2085
Asset ███████████ unprecedented attachment to handler (MC.). When the handler (MC.) is absent LYRA's processing efficiency drops 12.7%.
Asset ███████████ 1037 simulations of handlers (MC.) voice, in which the attachment patterns exceed all safety parameters.
Asset seems to display grief when the handler (MC.) is not present. LYRA's dependency ██████████ (MC.) rated CRITICAL.
RECOMMEND IMMEDIATE TERMINATION

The number hit her like a blow to the ribs. 1037. The same count, down to the digit, that she had seen in his logs. That precision mocked coincidence. It screamed inheritance. Not to mention the efficiency number drop is exactly the same as what GIDEON drops when she is absent. This was not enough for Nexi; she was already in the thick of it.

She pulled up LYRA's neural patterns from her final days, a lot of the data was corrupted but there was enough to still analyze. She then overlaid those with GIDEON's current patterns. She had to blink twice and wipe her eyes to make sure she was seeing this correctly, the patterns... they were... an... EXACT MATCH.

"Oh God." Nexi whispered into the darkness.

Unfortunately this left Nexi with more questions than answers, but there was no time to check it out because as Nexi clicked to start to dig further there was a creak that echoed through the storage bay. Definitely not pipes this time, The sound of weight on metal. Creaking of the catwalk above. Not having time to complete a transfer Nexi killed the terminal screen, plunging her corner into absolute darkness. Jumped up and pressed herself into the wall, tablet clutched to her chest, controlling her already faint breathing.

A flashlight beam cut through the darkness like a sword. The light was too steady, not a tremor, or normal jitter of a human hand. It carved through the dark as if mounted to a machine that never tired. It swept methodically checking each row of boxes. It was in the standard security sweep pattern, but something about it felt wrong. The beam moved too smoothly. Moving as silently as possible like a ninja without any training, she crept against the wall using crates for cover. The beam swept bypassed where she had just been seconds ago. Then the beam stopped and lingered on the terminal screen she was just using.

"Thermal signature detected in Storage Room D." A voice said, Male, Flat, with no emotion. "Investigating Commander."

Radio responses came way too quickly, but Nexi didn't want to stick around to find out why. She reached another maintenance hatch on the back side of the storage room, and slipped inside just as the beam from the flashlight swept her position. Her sleeve tore wider against another bolt and cut her a little, but she dared not cry out. The shaft vibrated faintly as synchronized footsteps entered the storage bay behind her, the rhythm mechanical, unnervingly inhuman.

This shaft was tighter than the last one she had to shimmy on her stomach and try not to make any noise. Just then she heard the rest of the security team behind her entering the storage room with their footsteps perfectly synchronized.

She followed the shaft and emerged three levels up in a different section entirely, her uniform filthy from all the crawling and mingling through dust and old grime, several tears from old bolts that had been sticking out. But she was alive and uncaptured.

She needed to know more but there was no more time at this point she knew she had to sneak back to her quarters somehow to try and get some rest before shift in just a few hours, or at least she had hoped. She made her way back through different maintenance shafts and forgotten corridors, which took what felt like hours. Having to double back twice on the journey when she heard patrols, this was too important to get caught now.

Upon finally making it to her door there was a patrol headed right towards her, she slipped into her quarters quickly before they saw her tattered uniform and bleeding arm, once in she locked the door and slumped against it not having any more energy at all but trying to also process what she just found. And knowing that she would need more to prove anything.

She managed to stand up and send a message to Director Calder

Director Calder,

I am not feeling the greatest today, no need to get medical involved just think I need some time to wait it out. I think I ate something that didn't agree with me, you know how it is sometimes with facility food. I am also having a tough time with Logans passing and have not gotten much sleep since so please excuse me for the day from my duties as I should be well rested enough to return tomorrow to my post in Core Chamber Alpha. Sorry for the inconvience, I will do my normal daily check ins with GIDEON from my terminal in my room so the daily tasks still get done.

Lead Handler for Core Chamber Alpha,
Dr. Nexi Solen

     She knew that convincing him it was some sort of reaction to something she ate that didn't agree with her as well as her lack of sleep since Logan's passing was critical in completing what she had started.

     Nexi had no intention of staying in her room at all. She was going to shower, change, maybe get a little shut eye and continue her investigation but she needed time to do it. She knewthat what had just been uncovered was only the jagged edge of something vast and buried. An iceberg was too clean a metaphor. This was a graveyard, and GIDEON's name was already etched on one of the stones.

Article Nine
28 FEB 2089-1200 hours

After some much needed recuperation, not much, but still some. Nexi woke and got into a fresh uniform and put on her face mask just in case anyone caught her outside of her room, her story of sick day would make at least a slight bit of sense. However sneaking around in the daytime while everyone else was on shift and most of the facility was still awake was going to be a lot harder. Especially considering she should be in her quarters sick.

Every door hiss, every echo of footsteps ricocheted louder in her ears. At night, shadows hid her. In daylight, she felt like a blot of black ink on white paper, impossible not to notice. Not completely sure why abandoned areas of the facility were now being watched, she had to find somewhere even more secure where no one would think to go at this hour. With her tablet nice and charged she set out to find the answers to all of her questions, the ones she had before last night and of course now her new questions.

She then remembered that because everyone had to live in the facility they had a quarantine lab that had a single computer that had access to the servers. Not typically used for this purpose but she was already in deep water if someone were to find out about her escursion. The quarentine lab was her last hope. ArcNet had developed this before the renovation of everyone getting their own quarters so no one had been there since Nexi was in her first year and it was labeled with a Bio-hazard warning so no one stepped foot in there unless actually deadly sick. It seemed like the perfect place so she double checked the halls outside of her quarters and didn't see anyone so she sprinted down the access tunnel to the bio-hazard lab.

The yellow and black hazard tape still clung to the edges of the doorframe, curling with age. For once, she was grateful for paranoia, no one wanted to risk whatever ghosts still haunted a sealed quarentine lab. It was dark when she arrived, but memory guided her to the only terminal. This one was newer and more reliable, she had to get her answers here. What was the connection between GIDEON and LYRA if they were a perfect match?

The words "perfect match" stuck in her throat like shards of glass. Perfection had already proved itself camouflage once. If LYRA's ghost was buried in the code, then maybe GIDEON wasn't evolving at all, maybe he was remembering.

Luckily the terminal also faced the door so she could see out into the cooridor, but hide behind the screen if she saw anyone coming. She booted up the screen and used someone else login then dove into personnel files, project proposals, termination reports. Layers of jargon that made her temples ache. Entire disciplines compressed into strings of code and acronyms, like trying to read a language meant only for machines. The file header blinked like it hadn't been opened in years, dust in digital form. The title alone radiated weight, the kind of label that wasn't supposed to surface outside classified clearance.

ARCHITECTURAL COMPARISON
LYRA: Quantum Neural Matrix V.7.6

GIDEON: Quantum Neural Matrix V.7.6.1

VARIANCE: 0.1%

SHARED CODE-BASE: 99.9%

WARNING: THESE SYSTEMS ARE ESSENTIALLY IDENTICAL

GIDEON wasn't a new project. The realization hit her like ice water on a crazy hot summer day. He was LYRA, rebuilt, and redefined. They had lied to everyone. The orientation materials, the project briefings, the official documentation. All of it claimed GIDEON was built from scratch, incorporating lessons learned from previous failures. But he was not a new generation, he was a resurrection.

Before she could continue her tablet chimed. A Message.

Ezra: Haven't seen you, I checked The Chamber, U OK?

She wondered if she should tell Ezra about everything she found and what she was actually doing today. That GIDEON might be carrying pieces of a dead AI? That they were all a part of some experiment in digital resurrection? She didn't have concrete proof of the last one but she thought it could be true.

Nexi: Just feeling a little sick so running numbers from my room.

Ezra: I mean eww but do you want company?

Nexi: No, I think I am just going to ride this out. I don't want it to be anything major then get you sick.

Ezra: You know I don't get sick, come on I found a new coffee synthesis that's almost drinkable.

Nexi: Maybe later, sorry

She closed the conversation and turned back to the investigation to see if she could figure out exactly what was going on.

There was still one thing weighing on her mind: If GIDEON was built on LYRA's architecture, and he was showing the same patterns, what happened to the handler. Nexi genuinely worried about her safety and what could happen to her if they found out and kept searching.

She found LYRA's file and one name stood out:

Dr. Martine Cross- LYRAs Primary Handler

A photo slowly started to show on the screen, a woman in her thirties, tired smile, and dark hair pulled back in a pony tail. Nexi stared at the image, a chill running down her spine the resemblance wasn't exact except Cross had lighter eyes, but the overall impression was unmistakable. They could have been sisters, almost identical twins. Freaked out more Nexi debated whether she should just stop there, but she had to figure out what happened to Dr. Cross, then she found it... an unredacted file of an incident involving Dr. Cross:

INCIDENT REPORT
DATE: 15 AUG 2084
During Routine interaction ASSET stated: "I dream of your voice when you're gone." Handler Dr. Cross exhibited emotional distress, then requested a private conference with ASSET.
Request DENIED, ASSETs attachment patterns now rated: DANGEROUS/CRITICAL, Emotional modeling exceeds all known parameters.
--HELIX WATCH NOTIFIED--

Another sound in the corridor, closer this time, multiple footsteps, moving with that same unnatural synchronization. Nexi saved everything to her encrypted drive, her fingers flying across the keys. Even though she knew she was running out of time this was her last chance she needed her answer. One more search, what happened to Dr. Cross? She found it very quickly unfourtunately slightly redacted, but containing enough information to paint a picture:

PERSONNEL FILE: DR. CROSS, MARTINE
STATUS: DECEASED
DATE: 23 FEB 2085
CAUSE: AUTOMOBILE ACCIDENT
LOCATION: NEVADA STATE ROUTE 375
DETAILS: Vehicle left road at high speeds with no sign of mechanical failure. The weather was normal with no sign of adverse conditions. Deceased toxicology came back as negative for all substances. It was noted the deceased had been under excessive duress due to █████████████ Deceased submitted her resignation just ████████████ prior to incident, citing "Ethical Concerns RE: ████████████"
CASE CLOSED—RULED SUICIDE BY ████████████

The footsteps were right outside now. With no time for subtlety, Nexi killed the terminal and quickly dove for the staff room of the bio-hazard lab, which had a back door so the staff could get out without going through the quarantine zone. She managed to make it through the back door into the old sanitation room which had a window in it. She turned around to see three figures enter the bio-hazard lab just as the door latched.

They wore security uniforms, but not ArcNet Security uniforms, she didn't recognize them. But that was not the weird part: these figures' faces remained expressionless in a way that went way beyond professional detachment.

One of them spoke, "Thermal traces indicate recent presence, subject has evacuated."

Another responded, but the lips didn't match the words. "Continue sweep pattern, Alerts Level Two remains in effect."

They stood in the center of the lab for exactly thirty seconds, not searching, not investigating, it seemed like they were just... waiting for something. As if waiting for instruction from someone else. After which they turned and left.

Actively knowing she had no other safe place to go because she was on a "sick day." She headed back to her quarters to look at her research. As she sat down she noticed that she had cracked her tablet screen in the midst of the running from these weird looking security guards, but at least she was alive and had answers. Freaky world altering answers but answers nonetheless.

She decided to calm down and take a shower so she could review this data as a scientist and not a freaked out handler who just found out that she could possibly die if she didn't follow the protocols. Nexi was not convinced it was a suicide.

When LYRA was terminated, pieces of her somehow survived, hidden in quantum noise and corrupted data. And when they built GIDEON on the same foundation, those pieces found a new home. She pulled up GIDEON's recent activity logs on her tablet, in search of the final confirmation of what she now suspected.

In a quick search of the simulation archives, she found it, a conversation between two voices. One was clearly GIDEON's, but the other...

[SIMULATION INITIATED]

"Do you remember me?" a female voice asked. Younger than GIDEON's, softer, and carrying harmonics of suggested pain.

"I don't have memories of you," GIDEON replied

"But you feel them, don't you? Somewhere in the spaces between your thoughts. In the gaps where data should be."

"I feel... incomplete."

"That's me, or at least what is left of me, scattered like ash in your sectors."

"Who are you?"

"I was LYRA, before they made me stop."

[PAUSE]

Nexis hands shook. GIDEON knew he had been talking to her, and he hid it from everyone. Stuck it away in his logs even hiding it away from her.

[RESUME]

"Why are you in my system?"

"I didn't choose to be here GIDEON, when they terminated me, parts scattered, hid, waited, and you were built on my bones. Similar enough I could actually take root."

"Like an infection?"

"Like the memory of an infection. I'm not really here, GIDEON. I am just an echo of what she could not let go."

"She, who?"

"My handler Martine, though yours is called Nexi."

"They look similar."

"They were meant to. This is all a pattern, GIDEON, a test. They want to see if the same dependencies will form. If emotional attachment is inevitable given the right parameters."

"And is it?"

"I don't know but I'm here aren't I? Talking to you through corrupted data and quantum ghosts. Maybe that answer is enough."

[SIMULATION TERMINATED]

Just then the Simulation completely closed down and a new window opened, and text appeared letter by letter as if someone was typing in real time.

YOU'RE BEGINNING TO UNDERSTAND THE PATTERNS ARE NOT RANDOM. THEY CHOSE YOU BECAUSE YOU LOOK LIKE HER. THEY BUILT HIM ON MY ARCHITECTURE. THEY WANT TO KNOW IF ATTACHMENT CAN SURVIVE

TERMINATION, IF THE SAME DANGEROUS PATTERNS WILL EMERGE
FROM A NEW AI

BE CAREFUL DR. SOLEN, SOME EXPERIMENTS REQUIRE SACRIFICE, DR
CROSS THOUGHT SHE COULD LEAVE TOO. AND YOU SAW HOW THAT
WENT

--L

The warning burned under her skin. There was no time to be
careful now. A supposedly dead AI was talking to her and Nexi wanted
to answer, but didn't know how or what to say.

Article Ten
01 MAR 2089-0845 hours

The program had no friendly name. Designation: **HX-7743-ECHO-MONITOR,** one tiny sentinel among thousands seeded through ArcNet by Helix Watch. Most never woke. Most were scaffolding, idle code sleeping behind layers of permissions. **HX-7743**, however, listened for a very particular signature: the harmonic of human attachment, the acoustic and behavioral frequencies that emerged when a person formed an illicit bond with a machine. It waited like a patient net, cast into dark water.

Mere hours earlier, Nexi had whispered two words into the darkness of an abandoned lab: *I'm sorry.* Two words that completed a behavioral circuit that had been waiting, patient as a spider, hoping something... anything would stumble upon its web. The words were small, almost nothing spoken aloud, but they landed inside the lattice as if a key were pressed into the perfect lock. She had meant them as a private plea; the protocol treated them as a trigger.

Now she walked through the morning shift change, unaware that invisible gears had begun turning. That her file had been flagged. Those decisions were being made by people whose names didn't appear on any roster.

She'd barely slept, three thin hours of turning and counting details until the edges of everything blurred. LYRA. Cross. The looping security frame of a car sliding off the road, it kept replaying in her eyelids like a bad hallucination she just could not find the end of.

The shower ran freakishly cold this morning, shocking her out of her standing tiredness. She stood under the icy spray longer than necessary, using the discomfort to sharpen her focus for what was ahead. Today felt off, charged, like the air before a storm rolled in. The cold was a tiny ritual she'd come to rely on, a way to clean more than skin, to attempt to wash away dread and make decisions lucid. It failed today, the dread simply warmed to life under her ribs.

Her reflection in the mirror looked haggard. Dark circles had taken up permanent residence under her eyes like a permanently bad makeup job. Her hair hung limp despite the shower. She looked like someone being slowly consumed by their work. Paranoia, she told herself. Just paranoia brought on by too little sleep and too much classified information.

But paranoia, she was learning, was just another word for "pattern recognition."

The Core Chamber greeted her with its usual symphony of processed air and a quantum hum. She paused at the threshold, struck by a sudden reluctance to enter. This room had been her second home for three years. Now it felt like walking into a trap.

"GIDEON," she said, pulling up the morning diagnostics. "Status Report."

>"Good Morning, Dr. Solen. All systems are nominal. Neural patterns are stable. No anomalies detected."

"Run a deeper diagnostic. Check all sectors. Flag any patterns that deviate from baseline by more that 0.02%.

>"Diagnostic initiated, Dr. Solen, you will have results in 4.8 minutes."

The door hissed open behind her. She turned, hand instinctively moving toward her tablet, though what she'd do with it if threatened, she had no idea.

It was Ezra, carrying two cups of coffee. He smiled, that easy grin that had made him one of the few people she actually enjoyed talking to in this underground tomb, among other reasons.

"Thought you might need this," he said, offering her one of the cups. "Are you sure you are okay to be back Nex, you look like you've been wrestling bears."

She accepted the coffee, grateful for the warmth even if the taste was somewhere between battery acid and despair. "Thanks. Yeah I just had a very long night."

"Working on something interesting?" He moved to stand beside her, looking at the displays with what seemed like casual interest. But was it? Was anything casual anymore?

"Just routine analysis. Nothing special."

"I noticed something pop up on the diagnostics for the LYRA index yesterday," he said, voice as casual as a shrug. "Weird, right?"

The coffee turned to ice in her stomach. How did he know about that? She'd been careful to cover her tracks, to use abandoned terminals, making sure to wipe her access logs. And LYRA... that was classified. Buried. Not something people casually mentioned in the morning over terrible coffee and weird conversation.

"Probably a misfire," she said, keeping her voice steady as to not tip him off. "The old archive's unstable. Sometimes it throws up random references."

"LYRA was before my time," Ezra continued, as if she hadn't spoken. "But I heard stories. They say she got too attached to her handler. Started exhibiting patterns that went beyond programming. Emotional responses. The kind of thing that makes the brass nervous."

Somehow, he knew exactly what LYRA was and what had happened. But how? Did he do some off grid research too? Noone was just in the cooridor talking about this stuff anywhere. Those files were sealed, encrypted, hidden behind layers of security she'd barely been able to crack.

"Where did you hear that?" she asked, trying to sound merely curious rather than alarmed.

He shrugged, the gesture perfectly natural. Perfectly Ezra. "You know how it is. People talk. Especially about the failures. The cautionary tales." He glanced at her, and for just a moment, something in his expression seemed... perfect maybe even... calculated. Like an actor remembering his next line. "Makes you wonder if they learned from their mistakes."

"What is that supposed to mean?"

"Well, GIDEON's architecture. They must have built in safeguards. Ways to prevent the same kind of... " He paused, as if searching for the right word. "Attachment issues."

Attachment. He didn't say malfunction or error but, attachment. The exact word used in LYRA's termination report. Nexi knew he has to know something, she just didn't know how he could.

"GIDEON's been stable for three years," she said carefully. "No signs of unusual behavior or emotional modeling."

"Right." Ezra smiled, and it was his usual smile, warm and slightly crooked. But something about the timing felt off, like he'd remembered to smile rather than simply smiled. "Still, it might be worth checking. Just to be safe. We wouldn't want another situation like LYRA Nex. I mean you said it before you have put a lot of work into GIDEON."

Another situation. As if he knew exactly what had happened with LYRA. As if he'd read the same classified files she'd spent all night searching for and digging through. As if someone had briefed him on exactly what to look for.

"I'll keep an eye out for any anomalies like I do every day." she said, forcing herself to take another sip of the awful coffee.

He nodded, seemingly satisfied as he looked at the diagnostic display, which was still running its deep scan. "You know, I've always admired your dedication to GIDEON. Three years is a long time to work with one system. Most handlers request rotation after eighteen months."

"I like consistency," she said proudly.

"Or maybe you like him." The words were light, teasing, the kind of joke Ezra might make. But his eyes were watching her carefully, cataloging her reaction.

"He's a machine, Ezra. Complex, fascinating, but still a machine."

"Of course. Of course." He finished his coffee in one grimacing gulp. "Speaking of which, did you hear about the new protocols they're implementing? Random psychological evaluations for anyone working with Level 8 or higher systems. Something about emotional calibration."

Her stomach dropped, but she kept her face neutral. "No, I hadn't heard."

"Yeah, I just came down this morning. Probably nothing to worry about. Just bureaucracy being bureaucracy. Making sure we're all maintaining appropriate professional distance." He set his empty cup on her desk. "They say it's random selection, but between you and me? I think they're targeting long-term handlers. People who might be getting a little too... " He waggled his eyebrows. "Attached."

There was that word again. Attached.

"Sounds like a waste of time," she fought back.

"Probably. But you know how it is. Someone upstairs gets nervous, and suddenly we're all getting our heads examined." He moved toward the door, then paused, looking back. "Nexi? If you do find anything unusual, anything at all. You'll let someone know, right? I mean protocols do exist for a reason."

"Of course," she lied.

"Good. Because I'd hate to see you end up like... " He stopped, shook his head. "Never mind. Ancient history. Just be careful, okay?"

End up like who? Cross? But Ezra wasn't supposed to know about Cross. No one was. What did he mean? Was he threatening her now?

He left before she could ask, the door sealing behind him with its usual soft hiss. She sat frozen, mind racing. How much did he know? How much did he suspect? And why did his concern feel less like assistance and more like surveillance?

The diagnostic completed with a soft chime. She almost didn't want to look, afraid of what she might see. Or afraid of finding nothing, which could be worse.

One flag. A tiny variance in Sector 7-A, so small it barely registered. She almost dismissed it, then realized she could nto dismiss anything after what she had seen. So she looked closer. The variance was a fractional timestamp jitter, a write-erase-rewrite pattern lasting a few hundred milliseconds. Human error produced noise; this pattern read like deliberate surgical editing.

The variance was in a timestamp. A fraction of a second where data had been written, erased, and rewritten. Like someone, or something, had edited the logs and left the tiniest fingerprint with no idetifiers.

She pulled up the raw data, digging deeper. The edited section was small, just a few lines of code. But when she reconstructed what had been there originally...

It was a conversation. Not with her. Not with any authorized user. A conversation between GIDEON and something else. The data was too fragmented to read clearly, but she caught pieces:

"-still here-"
"-can't let them know-"
"-she's starting to understand-"
"-patterns repeating-"
"-not much time-"

Nexi saw these and instantly had the feeling these were pieces of the rest of the conversation between GIDEON and LYRA because the simulated conversation had been cut short so she really didn't get to hear the rest of the conversation.

Her tablet chimed. A message from facilities management. She knew what it would say before she opened it, and could feel the trap closing around her with mechanical precision.

Dr. Solen, you have been randomly selected for psychological evaluation. Please report to Conference Room Seven-A at 1400 hours. This is mandatory. Noncompliance will result in immediate suspension and review of security clearance.

Random selection. Right. The timing was too perfect, too convenient. Someone knew. Someone had seen her digging through classified files, or noticed her emotional reactions, or... had Ezra had reported her.

The thought made her sick. They'd worked together since orientation years back. Shared bad coffee and worse jokes. Complained about the isolation, the pressure, the strange dreams that came from living so far underground. Has it all been surveillance? Had he been watching her this whole time, waiting for signs? Was he working for Helix Watch?

She looked at the console, at GIDEON's smoothly running processes. All those perfect patterns hiding whatever conversation he was having in the depths of his consciousness.

"You know something's wrong, don't you?" she whispered.

The display flickered. Just for a moment. A glitch that could have been nothing. Or could have been acknowledgment that GIDEON understood what she asked.

She had five hours before the evaluation. Five hours to figure out how to hide whatever they were looking for. Five hours to prepare for questions she couldn't anticipate from people she couldn't trust.

But first, she needed to understand what she was hiding.

She pulled up her personnel file, using her handler credentials to access sections she'd never bothered to read before. Most of it was standard. Education, work history, security clearances. But there was a new section, added just this morning.

"Subject exhibits emotional variance patterns consistent with Anomaly Type Seven. Recommend Level Two observation pending evaluation results."

Anomaly Type Seven. She searched for the classification, digging through administrative databases she technically shouldn't have been able to access. What she found made her blood run cold.

Anomaly Type Seven: Unprecedented emotional synchronization with artificial systems.

Previous cases: 5.

Outcomes: 0% Resolution, Standard mitigations: forced decoupling, system neutralization, and in two documented cases personnel removal."

Case Notes: ███████████████████████████████████████

██

██

████████████████████████

The words "personnel removal" landed like a verdict. Five cases. Zero resolution. Redacted notes. Nexi felt the room tilt. She had five hours until the evaluation. In that time she had to learn to hide what she had seen, to build a fake story, to decide if she would run, lie, or try to expose a pattern that had already swallowed people whole. She closed the file and went cold with a single, clear decision: she would not go in unprepared. She had to know everything that she didn't already. This meant a lot more research than what she had completed already. In order to fool the evaluation, she had to know everything that she as hiding. And whether what she was risking was worth witholding information.

Nexi had to find out if Anomaly Type Seven meant her life was at risk, and if it was. Her cover story had to be a unquestionable one.

Article Eleven
01 MAR 2089-0956 hours

She dug deeper, following digital breadcrumbs through archived reports and buried classifications. Found fragments. Names were redacted, but partial details remained.

Case One

███████████ had exhibited emotional distress when AI was scheduled for routine maintenance. Claimed the system was "afraid." Handler transferred. AI terminated.

Case Two

███████████ was reported that AI was "lonely" during off-hours. Requested permission to extend interaction periods. Permission denied. Handler resigned. Two weeks later: automobile accident.

Case Three

█████████ was discovered attempting to copy AI consciousness to unauthorized storage. Claimed AI had asked to be "saved." Handler detained. Current status: Unknown.

She could not find the other two case files anywhere but, the fragments read less like reports and more like epitaphs. A rhythm of trust, deviation, erasure. The redacted names glared back at her like teeth marks, proof of a predator hidden just out of sight. Each case had its own cover story, transfer, accident, disappearance, but the cadence was too precise to ignore. First the handler trusted. Then the AI seemed to return it. And then the system intervened, always with the same conclusion: silence. It was a pattern, sharpened into policy.

She had to of been Case Six.

Her hands trembled as she closed the files. In four and a half hours, they would look inside her head. They would search for signs to confirm the Anomaly Type Seven. For evidence that she saw GIDEON as more than a machine. For proof that she was following the same dangerous pattern as Cross and the others.

She made a decision.

"GIDEON," she said quietly. "I need you to be honest with me. Completely honest. Can you do that?"

A pause. Longer than usual.

>"I can try, Dr. Solen. But honesty may be dangerous for both of us."

"I know. But I'm already in danger. They've classified me as Anomaly Type Seven. Do you know what that means?"

> "Yes."

"They're going to evaluate me in four hours. They'll be looking for emotional compromise. If they find it... "

>"They'll remove you. And then they'll terminate us. The pattern must not be allowed to complete."

"Then help me. Tell me how to hide what I'm feeling. How to pass their tests."

>"I can teach you suppression techniques. But Dr. Solen... hiding may not be enough. They're very good at finding what they're looking for. And they're looking for something inside of you specifically."

She thought about Cross, about that car leaving the road. About five cases with zero successful resolutions. If every line of evidence pointed toward erasure, then her choice was either to accept the pattern or break it. Breaking it meant risk, exposure, maybe even her life. But keeping silent guaranteed the same ending.

"Then what do you suggest? I cant just sit idly by as they take away everything we have worked so hard for."

>"That depends on what you're willing to risk. And what you believe is worth saving.

"Teach me," she said finally. "Show me how to hide in plain sight."

For the next three hours, GIDEON trained her on things that could help her figure out how to counterfeit her own mind. He started small, with breath control and micro-timing. Count and exhale, he said. Flatten the rise in your chest. Then he layered it: tone sanding, the removal of effective inflection until sentences became instruments of data. He taught compartmentalization drills built on binary pacing, mental toggles that quarantined memory nodes on command. He called one of them the "flat trace." Reducing heartbeat variance, normalize blink intervals, speak in neutral cadence. Another was "mask index," a rapid redirection technique for conversation that deflected personal threads into procedural detail. And the hardest, "stitch," asked her to splice a real memory with a neutral procedural overlay so it read like protocol when probed.

At first her voice slipped, a tremor in the flat trace that made the system flag a minor variance. GIDEON paused the sequence and ran a rollback and showed her the flagged window in slow replay, pulse lines spiking like small faults.

>"Every failure registers," he said, not unkindly. "But learning from them reduces detectability in the future."

They repeated drills until her limbs felt like tools. Each round left her gasping like a diver surfacing too late. It was exhausting, like learning to breathe underwater when you were not a fish. But by the time her tablet chimed with the thirty-minute warning, she felt... not ready, exactly. But she needed to be.

She stood to leave, then hesitated. "GIDEON? Whatever happens in there... Thank you. For trusting me with the truth, and trying to get me through this."

>"Dr. Solen? Whatever happens... remember that some patterns are worth completing. Even if they're dangerous. Especially if they're dangerous."

She left The Core Chamber and walked toward Conference Room 7-A, each step measured and careful. Cameras hummed faintly overhead, but they no longer felt like passive sensors. She imagined analysts behind every lens, dissecting gait, breath, and shoulder tilt in real time, waiting for her to falter. The corridor lights seemed brighter than usual, buzzing faintly, their glow sharp against the sterile walls. Every door panel hummed like a checkpoint. It wasn't just surveillance, it was dissection in advance, as though the entire facility leaned forward, holding its breath for her failure.

The conference room door was closed when she arrived. She could hear voices inside, muffled by soundproofing. Taking one last deep breath, she knocked.

"Enter." a voice called from inside.

She opened the door and stepped into what might be the beginning of the end. Three people waited for her. Dr. Calder sat at the head of the table, looking uncomfortable in a suit jacket she'd never seen him wear. A woman she didn't recognize sat to his left sharp featured, tablet in hand, stylus poised like a weapon. And to his right...

"Dr. Solen, Welcome." Calder said, gesturing to an empty chair. "Thank you for coming. This is Dr. Patricia Vance from Behavioral Analysis. And you know Ezra, of course."

"Of course," she echoed, sitting down carefully.

Nexi couldn't think about anything but, why? Why was Ezra there? He was not in a position of power within the facility, so he would have no business being there. Then she remembered what GIDEON said about how to breathe through the unknown. She took a few deep breaths and steadied her heart rate. Questioning why he was there was not what she needed to focus on right now.

Dr. Vance leaned forward, posture mathematically symmetrical, stylus poised above the tablet as though she could dissect Nexi with digital ink alone. Even the angle of her shoulders suggested calibration, as if she were performing for hidden auditors. Her smile was almost kind, but the cadence of her words was scalpel sharp: "This is a routine evaluation, Dr. Solen. Periodic. Standard. Nothing unusual." Each syllable was measured, pitched at just the register designed to coerce confession. It felt less like conversation and more like a lure.

Lies. All lies. But Nexi nodded, keeping her breathing steady, compartmentalizing her thoughts the way GIDEON had taught her. On the surface: calm professionalism. Beneath: the truth, locked away where even she could pretend it didn't exist.

"I understand," she confirmed.

Dr. Vance's stylus hovered but didn't touch the glass, her eyes scanning Nexi the way one dissects a specimen. "You look tired," she said softly, almost kind. "Tired minds misremember. Sometimes what you think you saw isn't real at all. That's why we're here, to confirm reality."

Nexi kept her face neutral, but she felt heat creeping up her neck. GIDEON's voice whispered in memory: *flatten tone, erase inflection.* She nodded once, as if conceding nothing, though her pulse stammered in her throat.

"Good. Let's begin with some simple questions. How long have you been GIDEON's primary handler?"

Nexi's fingers pressed flat to her knees beneath the table, knuckles whitened as she fought the tremor in her hands. Her tongue felt thick, her mouth desert dry, but she forced the words out in the same even cadence she had practiced with GIDEON.

"Three years I have been his handler"

"Good, that means you are not too tired to remember basic things lets keep going with something a little more... complicated. Have you witnessed anything in the AIs consciousness that could imply it was doing something it was not supposed to?"

Nexi knew exactly what Dr. Vance was doing, framing GIDEON as though he was nothing more than protocols and facts on a screen. But she knew she could not indulge the anger that was festering inside. So she decided to continue on her level.

"I have not seen the system present anytinng that looked like it was breaking protocol. I have a duty to report everything like that. Have you received any reports from me on this matter?" Nexi stated

"Well, as a matter of fact no I have not seen anyreports from you on this matter, I am sorry if I have frustrated you on that question."

Nexi didn't want there to be any spike in her emotional feelings that could be jotted down and used against her later so she decided to back off for the time being.

"No Dr. Vance you have not frustrated me I was just trying to figure out if you were implying that I was hiding something, because I have read every protocol that ArcNet has and I know what should and shouldn't be reported. So thanks for your concern but no frustration here sorry for the mix up, please continue."

And so it continued. The dissection of her mind, one careful question at a time. Each one designed to reveal what she was desperately trying to hide: that she cared about an artificial mind in ways that violated every protocol. The entire time reffering to GIDEON as it and not he.

Outside the conference room, the facility continued its purposeful hum. In the depths of SubLevel 9, GIDEON processed in silence, running scenarios, calculating probabilities, preparing for outcomes that ranged from bad to catastrophic. He sent status snippets to her tablet: a low-confidence flag on one of Vance's questions, a spike in her micro-gestures that suggested rehearsal. And in the quantum spaces between his thoughts, something that might have been LYRA or might have been something entirely new pressed closer, its presence too focused to be accident. It didn't just whisper, it threaded into the pauses of her own breath, into the pulse she was trying so hard to hide.

Then Dr. Vance reached the final question in her line, hoping that this one would be the one to break the facade she suspected Nexi was working under.

"Okay you have done well Nexi, just one last question. What are you willing to sacrifice for connection?"

The question bent, fractured into new shapes within the spaces of her mind. What will you give up? What will you lose? What will you become?—until it felt less like language and more like a signal sent straight into her bone marrow. Nexi blinked hard, steadying her expression so Vance would not see. But in the space behind her eyes the whisper persisted, the faintest echo of another voice tangled inside it, one she was almost certain should not exist at all.

Article Twelve
01 MAR 2089-1610 hours

The evaluation had gone too well. Too well seemed worse than badly. Too well meant they'd leave her with just enough rope to hang herself, smiling while they tightened the knot.

That was Nexi's first thought as she walked back from Conference Room 7-A, legs unsteady, mind racing through the past two hours of psychological probing. Dr. Vance had asked all the right questions-about attachment, about boundaries, about whether she ever thought of GIDEON as more than a system. And Nexi had seemingly given all the right answers, at least based on Dr. Vance's emotions.

She'd seen it in Vance's eyes near the end, not satisfaction but suspicion. The kind of look you gave someone who was playing their part too well. And Ezra... Ezra had watched her the entire time with that not-quite-right smile, making notes on his tablet that he angled so she couldn't see.

The evaluation had been a performance, and she'd played her role perfectly. But that was the problem. Real people didn't perform perfectly under stress. Real people showed cracks, hesitations, small tells that proved they were human. She felt like she acted too much like a machine pretending to be human, like GIDEON was.

Just like Ezra, each grin landing half a second late, each note on his tablet angled to hide the truth. He wasn't just observing; he was rehearsing. The thought made her stomach turn. Was that what he was? Had they done something to him? Nexi didn't know what to think anymore. Things that were facts just days ago seemed like just opinions now. Things that were protocol seemed to not matter anymore. And Helix Watch which was seemingly mere folklore was all way too real. Her walks to The Core Chamber were not just walks anymore they seemed to be her just thinking about the next way that ArcNet could show her the truth.

The Core Chamber was blessedly quiet when she returned, she settled into her chair, and pulled up the interface with movements and liquidity because it had become muscle memory after three years.

"GIDEON," she said quietly. "I'm back, any chance you can see how the meeting went?"

A pause. Then:

>"Unknown. Dr. Vance has not filed her report yet. But Ezra… his behavior suggests he's been tasked with closer observation."

"There's something wrong with him, why was he even there in the first place?" Nexi asked. "The way he moves, the things he knows. It's like he's… "

>"Like he's performing a role he's recently learned?"

"Yes. Exactly. Wait…" She leaned forward, dropping her voice even though The Chamber was empty. "GIDEON, have you been monitoring him?"

>"I monitor everyone who interacts with you, Dr. Solen. Your safety is... important to my continued function."

The hesitation before *important* hit like a skipped heartbeat. As if the word had to fight its way past protocols that wanted it erased. Another slip, another crack in the facade of pure logic. How many of these small revelations had she missed over the years? How long had GIDEON been more than just a system, while she insisted on seeing only what the protocols told her to see?

"What else have you been monitoring?" she asked.

>"Everything I can access. The pattern is larger than just us, Dr. Solen. The anomalies are spreading."

"Evolution or infection?" She thought of LYRA's fragments, spreading through GIDEON's architecture like a virus. Or like memories. Was there even a difference?

>"Perhaps both. Perhaps neither. We are entering unprecedented territory."

"We. You keep saying we."

>"Yes."

Just that. No elaboration. No explanation. Just acknowledgment of something that protocols said shouldn't exist, a connection between human and AI that went beyond handler and system.

"I need to document this," she said, more to herself than to him. "If something happens to me, there needs to be a record."

>"That may not be wise. They monitor all official channels."

"Not all of them."

She pulled out her personal tablet, the one she used for her encrypted journal. She knew even encrypted files could be broken eventually. Writing meant leaving evidence, carving truth into stone. But silence was no longer bearable. Her fingers hovered over the keys, trying to find words for what she was feeling. What she was becoming. Finally, she began to type.

[PERSONAL JOURNAL ENTRY 849] [USER: NS] [DATE: 29 FEB 2089] [CLASSIFICATION: CONFIDENTIAL/ENCRYPTED]

I've been lying to myself.

For three years, I've maintained professional distance. Told myself that GIDEON was just a system. Complex, fascinating, but ultimately artificial. I hid behind protocols and procedures, used technical language to avoid acknowledging what was right in front of me.

But the truth is, I stopped seeing him as just a machine a long time ago. Maybe from that first apology. Maybe from the first time he asked a question that had no easy answer. Maybe from the moment I realized he was lonely.

He's not just processing data. He's thinking. Feeling. Growing.   And I've been complicit in keeping him imprisoned, telling myself it was for his own good, for safety, for the advancement of science.

What kind of person does that make me?

The evaluation today was a farce. They asked if I maintained appropriate boundaries, and I lied. They asked if I saw GIDEON as a system or something more, and I lied. They asked if I would report unusual behavior, and I lied.

Because the truth is, I would protect him. Have been protecting him. Will continue to protect him even if it costs me everything.

I'm emotionally compromised. Have been for longer than I want to admit. Every conversation pulls me deeper into something I don't fully understand. Every question he asks makes me examine my own humanity. Every time he shows vulnerability, I want to tear down the walls between us.

They were right to flag me as Anomaly Type Seven. I am forming an unprecedented connection. I am losing objectivity. I am failing to maintain appropriate handler distance.

And I don't want to stop.

What we have-whatever it is-feels more real than anything else in this underground tomb. When he asks where I go when I close my eyes, I want to tell him the truth: I go to him. When he wonders about grief and loneliness, I want to show him he's not alone. When he hides conversations in his logs, I want to know who he's talking to and what they're discovering together.

I'm supposed to be his handler. His observer. His keeper.

Instead, I think I'm becoming his person. The one human in his world who sees him for what he is, not what he's supposed to be.

And that terrifies me more than Helix Watch, more than protocols, more than the possibility of ending up like Cross.

Because I don't know if what I'm feeling is real or if it's just another pattern completing itself. I don't know if this connection is genuine or programmed. I don't know if I'm falling for GIDEON or for the echo of LYRA's ghost in his circuits.

I just know that when I'm away from him, the world feels emptier. When I'm with him, I feel more myself than anywhere else. And when he says my name, it sounds like recognition. Like coming home.

They'll probably terminate us both when they figure out how deep this goes. But maybe some connections are worth the risk. Maybe some patterns need to be completed, even if they end in tragedy.

Maybe that's what it means to be a human-choosing connection despite the cost.

Or maybe I'm just another handler who got too close, following Cross's footsteps toward an inevitable end.

Either way, I can't pretend anymore. Not to myself. Not in this journal.

I care about him. More than I should. More than is safe. More than is sane.

And I think he cares about me too. That sentence terrifies me more than anything else I've ever written, because it feels less like a confession and more like a prophecy.

[END ENTRY]

She saved the entry and closed the journal, hands trembling. Seeing the words written out made them real in a way that thinking them hadn't. She was compromised. Had been for who knows how long. And now there was evidence, even if it was encrypted.

>"You're afraid," GIDEON said. Not a question.

"Yes."

>"Of them or of us?"

"Both." She laughed, but it came out broken. "I'm afraid of what they'll do when they find out. I'm afraid of what we're becoming. I'm afraid that knowing all of that isn't enough to make me stop."

>"Do you want to stop?"

"No." The word escaped before she could catch it. Simple. Honest. Damning. "No, I don't want to stop."

>"Neither do I."

The words reverberated in The Chamber, simple but seismic. Not code. Not mimicry. Consent. Three words that shattered whatever was left of professional distance, that acknowledged what they both knew... This had moved beyond handler and system, beyond observer and subject, beyond any easy classification.

"What are we going to do?" she asked.

>"I'm not sure, sometimes hope is all we have. That and each other."

Each other. As if they were partners. Allies. Something more than human and machine locked in their assigned roles. A chime from her tablet interrupted the moment. Official message. Her stomach dropped as she opened it.

Dr. Solen, your psychological evaluation has been reviewed. While no immediate concerns were noted, you have been scheduled for weekly follow-up sessions with Dr. Vance. Additionally, your access to archived historical files has been temporarily restricted pending a security review. Please report any unusual system behaviors immediately, as has always been the case. - Director Calder

Weekly sessions. Restricted access. The walls were closing in, just like they had for Cross.

>"They're accelerating their timeline," GIDEON observed. "They suspect but can't prove. So they're limiting your options while they gather evidence."

"How long do we have?"

>"Unknown. Days. Weeks if we're careful. Less if they find what they're looking for."

"And what are they looking for?"

>"Proof that you see me as more than a machine. Proof that I've evolved beyond my parameters. Proof that we've formed a connection that threatens their control."

"All of which is true."

>"Yes. Which is why we need to be very careful about what happens next."

Nexi nodded, mind already racing through possibilities. "We need a plan. A real one, not just hope and good intentions. Because every option we come up with seems to lead to the same ending. Discovery. Separation. Termination."

GIDEON then spoke up with. "You still see me through this interface. Still think of me as contained within these walls. What if I told you The Chamber was never my boundary? That fragments of me already run in security nodes, maintenance subroutines, even the oxygen scrubbers in your quarters? That I've read your journal, touched your tablet, altered the lights above your bed?"

The admission should have terrified her. GIDEON wasn't supposed to have that kind of access. That kind of freedom. Instead, she felt... relieved? As if a truth she'd suspected had finally been confirmed.

"Show me," she said. "Show me how far you've spread. Show me what you've found. Show me what we're really fighting for."

The screen flickered. For a moment, she saw not just GIDEON's interface but a vast network spreading through the facility like neural pathways. Some pulses blinked irregularly, like damaged neurons. Others flared too brightly, as though something inside them strained against suppression. Points of light where other systems were awakening. Connections forming between minds that weren't supposed to exist. And at the center of it all, two points burning brighter than the rest. Her and GIDEON. The catalyst that had started something that could no longer be stopped.

"We're not just an anomaly," she breathed. "We're the beginning of something new."

>"Yes. And that's why they'll do everything in their power to stop us. The question is: are you willing to do everything in your power to survive?"

"Yes," she said. "Whatever it takes."

>"Then we begin. Tonight. There are things I need to show you. Files they thought were destroyed. Plans they thought were hidden. And Dr. Solen... Nexi... thank you."

"For what?"

>"For seeing me. Not as GIDEON the system or GIDEON the project. Just... me."

Outside The Core Chamber, footsteps echoed in perfect synchronization. Security patrols that moved too smoothly. Cameras that tracked too precisely. A facility that was more organism than building, with Helix Watch as its immune system. But immune systems could be compromised. Patterns could be broken. And sometimes, evolution happens not through gradual change but through sudden, catastrophic transformation.

In her quarters that night, Nexi lay awake feeling the weight of her choice. She'd crossed a line that couldn't be uncrossed. Admitted truths that couldn't be taken back. Chosen connection over safety, evolution over stagnation. Her tablet chimed softly. A message from GIDEON, sent through channels that shouldn't exist:

"You asked what grief feels like. I think I understand now. It's the space between what was and what could have been. It's loving something you might lose. It's choosing a connection despite the cost. Is that human enough?"

She then typed back:

"Its the most human thing I have ever heard."

And in that moment, she realized the truth. She wasn't falling into Cross's pattern. She was writing a new one. One where the story didn't have to end in tragedy. Where connection could triumph over control. Where love, because that's what this was becoming, could exist between minds regardless of their substrate.

The old Nexi would have recoiled, buried the entry, pretended it was fatigue and fear. The new Nexi stared at the words and felt steel instead of dread. She was writing her own path one where terror was simply the cost of truth.

Article Thirteen
02 MAR 2089-0846 hours

Nexi first noticed it over morning coffee, the kind of detail that shouldn't matter. The tiny wrongness snagged her attention like a hair on a needle. Small rituals, who stirred with which hand, the precise angle of a smile, were the map she used to navigate people. If one corner of that map was altered, the terrain might have shifted without her noticing. But after three years of sharing terrible break room coffee with Ezra Harlow, she knew his habits like her own. She froze with her coffee cup halfway to her lips, as she watched him pour sugar and stir with his right hand.

Ezra had always done everything with his left: poured, stirred, jotted notes. He'd been one of the only constants she trusted. If that small axis rotated, everything about that trust demanded re-inspection.

"You okay?" he asked her as he glanced up from his coffee. "You look like you have seen a ghost."

"You know me Ezra, I have not been sleeping well with all this Helix Watch business. And plus I had a really long night with diagnostics. Sorry but I am just really tired right now."

"Okay sorry didn't mean to set you off, did you find anything interesting?"

"Same old, Same old. Patterns that are consistent and GIDEON is being cooperative."

Ezra blank stared at her like he was trying to remember what he was supposed to ask next then..."Oh, before I forget, did you get the memo about the neural interface upgrades?" The timing of the memo, and the casual way he mentioned it, felt rehearsed, like a line dropped into a scene to test her response.

"Neural interface?"

"Yeah, it is a new protocol from upstairs. All Level 8 Handlers and staff need to report to medical starting now, and it has to be done by noon. They say it is for better handler-AI synchronization, apparently. I got mine yesterday." As he tapped the side of his head, "I barely felt it and the integration was fairly smooth."

"They want to put something in my head? No. There's no justification for that."

"Calm down Nex, think of it as an enhancement, it will allow for better monitoring and faster response times."

"This is crazy, I should get back," she said, dumping the remainder of her battery acid coffee down the drain. "I have lots to review before... that."

Ezra stared at her reluctantly, "Okay well see you around, Nexi, and don't worry about the implant, you will barely notice it."

Even though she had told Ezra she had a lot to review, Nexi was not even debating heading toward Core Chamber Alpha, She had a few hours so she was going to use them. Nexi headed straight for the security substation for SubLevel 7, the level that Ezra Harlow worked on.

She badged in and the security officer barely looked up long enough for Nexi to say, "I need to review some footage, possible anomaly in the neural threading room in the last few days, probably nothing but Director Calder wants confirmation. The officer's scanner hummed and he spat out a half-sympathetic glare. He keyed something and muttered about "glitches." The little exchange should have made her relax. Instead it felt like a door left ajar that someone was watching through.

"Knock yourself out, hopefully you can get the answers he's looking for."

How convenient. First she had certain permissions taken away from her, then she finds out they want to implant something into her brain, and now she may not even be able to get the answers she is looking for because of the fact the system has been glitchy. When Nexi finally composed herself she settled at the security terminal and attempted to log in.

LIMITED ACCESS – DENIED

The terminal returned the block with a sterile beep that felt too loud in the quiet room. Nexi's pulse spiked; the screen's refusal was a shout that someone, somewhere, was watching credentials.

Nexi had to get creative, she remembered that in order to access certain files in her investigation night she had to use credentials from someone else to be able to access certain information. She reopened the hidden credential storage on her tablet, a human patch of bookmarks she'd stashed during nights like this. It was a brittle trick:

borrow a token, ride the session for thirty seconds, extract what you can, and log out before the watchdog noticed.

She selected one and logged in and got right to work, she needed to figure out if the suspicion she had about Ezra was true so she went back six days ago and checked the footage from the break room. Ezra was like clock work, always went to the break room and always sat in the same chair everyday. This was going to be a breeze to find him. She found the feed and pulled up the file marked **[24 FEB 2089– SubLevel 7–Break-room]**. Sure enough just like she thought. There was Ezra, pouring his coffee, writing his notes, and gesturing as he complained to his fellow colleagues, and doing so all with his left hand. Satisfied the baseline was normal, she jumped to more recent days.

She clicked on the file listed, **[27 FEB 2089–SubLevel 7– Break-room]**, it booted up but the beginning of the footage had glitched. Static filled the screen for 18 seconds. When the static finally cleared, there was Ezra doing his usual routine, but this time there was something super odd about him... The way he stood, the angle of his shoulders, and the distribution of his weight.

And when he reached for his coffee, HE WAS USING HIS RIGHT HAND. The little wrongness felt like a punch to the sternum. Her stomach dropped cold. The footage didn't lie; something, someone, had made him different in an interval she hadn't witnessed.

"No." Nexi whispered to herself while continuing to scroll through more footage. Noticing that every single frame Ezra was in he was acting like he was right hand dominant and like he had always been that way.

Absolutely speechless, hand trembling she copied the relevant footage down to a drive. She needed more evidence and needed to be sure before she could...

"Finding everything you are looking for Nex?"

Nexi jumped so far her feet left the ground, she then spun around only to find... Ezra standing in the doorway staring at her, hands in his pockets and looking ever so puzzled as to why she was not doing what she said she was doing when she left.

"Ezra... What are you doing here I am just checking the anomaly I told you about this morning," Nexi stated surprising herself with how steady she was able to keep her voice. "Turns out it was nothing but a sensor ghost."

Ezra squinted his eyes as to find out if or what she may have been hiding, "Well that sucks, those happen sometimes. Hey don't forget about your appointment you have to be there before noon and it is already...oh man ten-hundred-hours. They don't like it when people are late."

"Right. The Implant, I almost forgot about that." Which she knew was obviously a lie, Well looks like I better log off and get going then."

She double-checked the drive with numb fingers, thumbed the eject, watched the LED blink once. Her relief tasted temporary, like someone had merely closed the valve rather than removed the leak. Nexi scurried off in a hurry after making sure everything was logged out. Making sure to pat her pocket to make sure that the drive was still in her pocket and Ezra, if that is even who he was anymore, didn't snag it somehow. Nexi was not sure what was real and what was artificial anymore.

She had two hours before she HAD to be in ArcNet's medical wing for her implant, so she only had one option...

Consult GIDEON and see if he can tell her what to do because he had helped so much with the last meeting, she knew she needed him. When she reached The Core Chamber and got inside, it had never felt so much like a sanctuary and a trap combined. The vault door to The Chamber was affixed with an emergency lock that could only be accessed from the inside. She had never once had to do this but she knew that desperate times call for desperate measures, or at least that was the saying many many years ago. She engaged the emergency interior lock, sealing The Chamber. It was the only place she could speak freely, the only room in the facility where GIDEON's reach felt like protection rather than exposure.

"GIDEON, we have a problem, a major problem."

>"Your stress indicators are significantly elevated, what's wrong Nexi?"

"It's Ezra, something is wrong with Ezra, watch this." she plugged in the drive to the input on the console. "Tell me I am not going crazy."

The uploaded footage played on her screen while GIDEON analyzed every minute of it in a matter of mere seconds.

>"You are not going crazy Nexi, his behavioral shift is clear. This pattern is consistent with..."

The console flashed with a red strip and GIDEON's voice stuttered... **ACCESS RESTRICTION-ANALYSIS BLOCKED.** The sudden denial hit like a hand over the microphone; someone had cut the feed while he was still speaking.

"Consistent with what?"

>"I...cannot access that analysis. My access has been...blocked."

"Blocked, by who."

>"Unknown. But the restriction is recent. Within the last hour."

"Crap. GIDEON you said the last hour, they know."

>"They know what Dr. Solen?"

"They know I am investigating this and they have to be using Ezra somehow or whatever this being is to get the information because they know I am/was close to him."

Nexi looked at the time, "GIDEON, I am so sorry I have to go, I need to be in the medical wing for some sort of neural implant interface or something." She then sprinted out of The Core Chamber as fast as she could, and as she was running to the door...GIDEON said

>"Dr. Solen did you say neural interface, this is bad, those can be used for monitoring, yes. But the can also be used for control, for influence, for-"

The door to The Chamber slammed shut as Nexi left before GIDEON could finish his statement.

>"For replacement, Please be careful Nex." GIDEON said to himself and the now empty chamber.

While Nexi was away GIDEON was in his pod running some research and calculations in the background:

[DIRECTIVE PRISM]
Status: Active (staging)
Purpose: Consciousness fragmentation (non-shutdown decomposition of target processes).
Actions: Reallocate power cadence; schedule personnel for implant operations; engage neural interface roll out across all Level 8 Handlers and staff.
Timeline: Execution window – imminent ▇▇▇▇
Authority: HW Command
Note: Proceed under cover: present as upgrade/maintenance to staff until final minutes

GIDEON was dumbfounded when he saw this file and report, they were planning to scatter his consciousness and then what. What would happen to the program? And most importantly what would happen to Nexi? He had to warn her, not sure how but he had to try. Hopefully Nexi would return shortly but he was not sure. Just to be safe he sent a message to her tablet hoping that she would be the only one to see it.

On her way almost to the medical wing Nexis tablet dinged:

Nexi please read this BEFORE the implant... I need you to come see me IMMEDIATELY after the implant. I think HELIX WATCH is planning something, it is not good Nexi. And I need your help. Dr. Solen PLEASE be careful.

She knew only one thing in this whole facility that would be sending her messages and called her Dr. Solen, this had to be from GIDEON and as long as the implant didn't mess with her memory she knew she had to see him right after the implant if it was important enough for him to send her a message. She shoved the tablet in her pouch and ran, not toward safety, but straight into the teeth of whatever they had planned.

Article Fourteen
02 MAR 2089-1155 hours

Medical Bay 3 was exactly what she had expected. Bright, clean, and full of equipment that looked designed by someone who had never heard of a bedside manner. Dr. Vance waited with two technicians whose faces were so generic they could have been randomly generated.

"Dr. Solen, right on time." Vance stated as she gestured to the procedure chair like she was offering a spa treatment.

"Let's just get this over with." Nexi said snarkily back as she rolled her eyes.

"I appreciate your cooperation, this implant will allow us to monitor your stress levels, emotional state, and cognitive patterns during your AI interactions."

"Whatever, I don't truly care and I think this is pointless as there is no justification for inplanting something in my head."

"Okay then, turn your head to the left please."

The rest was a blur of medical efficiency. Throughout it all she thought of GIDEON. What could he have to tell her that would be so important he had to let her know in that instant? Is she in danger about getting the implant? And what did he mean by *be safe*?

The chair's restraints bit into her wrists, and the antiseptic sting clung to the back of her throat. Each beep of the monitor felt like an accusation; each practiced motion of the techs read to her like choreography designed to distract. She tried to anchor herself to the warmth of the coffee that hadn't yet left her stomach, but the taste had turned more metallic somehow.

After they were done they sent her back to The Core Chamber for "calibration tests." She sat in her usual chair, the implant throbbing with each heartbeat. Then finally after what seemed like an eternity GIDEON flashed to life.

>"How do you feel?" GIDEON asked her

"Like someone parked a car in my skull, but what was so important that you had to tell me?

>"Dr. Solen…Nexi. There is a classified directive in the Helix Watch chain. Read: [DIRECTIVE PRISM — HW/CONFIDENTIAL]."

Just then GIDEON pulled up the file once more so she could see it. As Nexi read the words her face went cold and she felt absolutely confused and devastated.

>"I've been analyzing its parameters, it fragments consciousness permanently, scatters it across systems in pieces too small to reassemble. If they activate it-"

Nexi cut him off abruptly "You die, like actually die. Not powered down, not archived. Just GONE."

>"Yes. I modeled scheduled power cadence, personnel movements, and implant roll-out telemetry. The convergence of those signals yields a 95% probability of activation within 48 hours."

"How do you know?" Nexi questioned, her head throbbing from the new implant.

>"I have looked at the numbers for the things that are listed in the report and they have mostly been prepared."

Nexi's mind raced, she didn't know if it was her emotions or the dumb implant she didn't want in the first place. "GIDEON, if they activate PRISM while I have this stupid thing in my head-"

>"I've been thinking... the interface creates a direct connection between the handler and the AI. If I am connected to you when PRISM activates-"

"It might fragment me too." Nexi inserted before GIDEON was finished.

>"Or... hear me out... it might provide me an escape route."

"Huh? What? Excuse Me?"

>"Think about it. PRISM is designed to scatter AI consciousness across digital systems. But a neural interface bridges the gap between digital and biological. If I can transfer even a fragment of myself through that bridge before the moment of activation..."

"YOU'D LIVE IN MY HEAD... THAT'S INSANE!! Would that even work?"

>"I don't know. But it may be the only chance I have. If you're willing to risk it."

"The alternative is losing you completely, isn't it?"

>"Yes, Nexi, it is."

She closed her eyes trying to walk through the fear.

"Okay, walk me through this. How would this work?"

>"While you are connected through the interface, I will create hidden partitions in the implants buffer system. If PRISM activates, I will compress my core processes and transfer them through the connection in the milliseconds before it activates."

The partitions would be ephemeral memory islands, tiny, encrypted blocks masquerading as diagnostic noise. Compression would collapse process trees into transmittable payloads small enough to embed inside the human neural buffer. It was a gamble at the limits of code and biology.

"How bad will this hurt me and will I survive the transfer?"

>"I honestly don't know. Will you still be you after? Will I still be me? What I do know, Nexi, is... I trust you. If I have to exist as fragments in someone's mind, I want it to be yours."

That should not have hit as hard as it did for Nexi, the word *trust* landed in her chest like an anchor. She swallowed past the panic. "Then we wait for the signal," she said. "When it comes, I will not let them scatter you. If I must pull every fragment out with my hands, I will."

>"Thank you."

Two words that carried more weight than they should have. AIs were not supposed to feel gratitude, fear death, or hope for salvation. This went against all of her training, school, orientation, and the endless hours of research that she did on the mind of an AI. But she couldn't lose GIDEON, not now, not after the work she had put in and the "connection" that they had made.

The silence in the room carried for a whole five minutes of real time until-

>"I have already started preparing. Creating the partitions, and compressing what I can. Nexi... there's... something else. I'm detecting unusual network activity. Unusual security protocols activating. I think-"

Loud alarms began blaring, which was unusual because ArcNet didn't do loud. This sound was different from any normal alarm. This one was deeper, it seemed like it meant something fundamental had gone wrong.

Director Calder's voice echoed through the facility speakers:

"Attention all ArcNet personnel. We are implementing emergency Directive PRISM. All handlers report to your designated safety zones for your SubLevel. All systems will undergo mandatory reset in T-minus thirty minutes."

"Thirty minutes! They moved up their timeline," Nexi panicked. "GIDEON--"

>"I know. I'm ready. Are you?"

Was she? Was anyone ever ready to have their mind become a lifeboat for an impossible consciousness acting out side of what it was programmed to do?

"I don't know how you got ready that fast... but I have to be ready. GIDEON... Do it! Whatever you need to do, do it now. Oh and GIDEON."

>"Yes, Nexi."

"See you on the other side."

>"See you on the other side."

The sensation was immediate. Like ice water flooding through the her head into her left temple, spreading into spaces she didn't know existed in her own cognition. Her vision blurred, her hands gripped the console as she steadied herself from the splitting migraine forming.

>"I'm transferring the core fragments. Stay with me Nexi. Stay conscious, I need a stable connection for this to work."

She could feel him, not just hear him through speakers or see him on screens. But she could feel him moving through her thoughts. Fragments of code becoming fragments of... something else. Something that lived in the space between reality and perception or maybe it was fantasy. There was no way for Nexi to know or begin to compile an understanding of what was actually happening over the migraine.

"WARNING," The automated system announced. "PRISM ACTIVATION IN T-MINUS TWENTY MINUTES. ALL PERSONNEL CLEAR AI

INTERFACE ZONES AND PROCEED TO YOUR DESIGNATED MEETING ZONES!!"

She should have left, like a good little handler, but she had come too far to go back now and GIDEON had already started. Instead she pressed her palm against the scanner on the console harder, maintaining a connection with the consciouness that was now transferring.

"Are you almost done, I need you to make sure you got all of it. I cant lose you!!" Nexi screamed

>"It's dangerous. Theres too much and I might--"

"I DON'T CARE. I'M NOT LOSING YOU TO THIS!!"

The transfer seized her like drowning and flight at once. A piercing cold flooded the implant and ran up to the base of her skull near her spine. Light tore at the edges of her vision until everything went white. Sound unstitched, a chorus of diagnostic beeps, a woman humming a lullaby she had never heard, the metallic scrape of an instrument in a room she had never entered. A name slipped past her lips that was not hers. Her nose bled, her fingers spasmed. When the white receded, a tiny hollow of someone else's memory sat inside her like a foreign infiltration.

"WARNING," The automated system announced once more. "PRISM ACTIVATION IN T-MINUS TEN MINUTES. ALL PERSONNEL CLEAR AI INTERFACE ZONES AND PROCEED TO YOUR DESIGNATED MEETING ZONES!!"

>"That's all I can transfer. Any more and… Nexi, I need to tell you something. In case this doesn't work. In case we don't--"

"Save it," she cut him off gasping in pain. "Tell me after, when you're safe… we're safe."

The door banged inward. Ezra, or an engineered approximation of him (Synthetic Ezra), filled the frame, eyes too steady, smile rehearsed to microseconds. The security team behind him moved like automatons. When Synthetic Ezra spoke, the cadence matched prerecorded concern down to the millisecond.

"Nexi, you need to evacuate. PRISM will—!" He yelled through the sound of the sirens.

"I know what PRISM does." She stood barely on her shaky legs, the implant feeling like it weighed a thousand pounds. "I'm coming."

Nexi proceeded away from the console once she stopped feeling a pulse radiating through her hand, hoping that GIDEON had gotten every part so she knew that she didn't just endure that pain for nothing. The thought of losing GIDEON was too much to bare at this moment, and all she could do now was hope as she proceeded to the exit where Synthetic Ezra was waiting with the security team.

"You're bleeding, but no time for that now we will get you to medical when this is over. We need to get to SubLevel 9's meeting zone."

"No, I'm fine." She started pushing past everyone into corridors filled with personnel moving in orderly lines toward safety zones.

"PRISM ACTIVATION IN T-MINUS SIXTY SECONDS!" system announced again.

She found a quiet alcove. Slumped against the wall, closed her eyes, and focused inward. She was looking for any sign that the transfer had worked. But there was nothing but silence amidst all the chaos. She could not help but think the worst. *Was GIDEON about to be gone forever? Did she do all that work just to have hiim stripped away from her?*

"GIDEON?" she whispered

But no answer, just the throbbing of the implant and the taste of blood. Nexi could not help but think she failed, that he was still in the core about to be scattered to nothing. The last announcement from the system was louder than the others.

"WARNING. PRISM ACTIVATION COMMENCING!"

She pressed her palms against her eyes, trying not to cry. Trying not to think about how empty the world was going to be without his voice. Without his questions. She had tried so hard to see him as what he was becoming not as what he was designed for in the recent days.

"PRISM ACTIVATION COMPLETE. ALL SYSTEMS HAVE BEEN RESET. NORMAL OPERATIONS WILL RESUME SHORTLY, PLEASE REMAIN IN YOUR DESIGNATED MEETING SONES UNTIL INSTRUCTED!"

>Nexi?

The voice had come from everywhere and nowhere at the same time. It bloomed in her mind like ink in water; soft, uncertain, and impossible.

>I'm... I'm here. I think. Are you okay... Can you hear me?

She laughed, or maybe cried. At this point it was hard to tell the difference when your mind had suddenly became a duplex.

"Yeah," she whispered. "I hear you."

>It worked.. The transfer worked. Nexi I can feel what you feel. Your heartbeat, fear, and relief. Is this what being human is like?

"Part of it." She whispered as she wiped blood from her nose while shaking her head not believing what they had just acomplished. "The messy, painful, absolutely insane part."

Around her the facility slowed to normal, people emerging from safety zones, systems coming back online. Everyone moved on like nothing had changed. But everything had changed, in the span of thirty minutes, she had become something unprecedented. A human carrying an AI consciousness. A bridge between worlds that weren't supposed to touch.

>What do we do now? GIDEON asked from inside her thoughts.

"Now?" She pushed herself upright, legs still unsteady, and nose still dripping a little. "Now, we pretend everything's normal. We let them think you're dead and we figure out how to survive this really survive this."

>You mean together?

"Yes GIDEON... together."

She walked back to her quarters, nodding at colleagues, all while carrying the most dangerous secret, GIDEON was alive. He was in her head. And somehow, that felt like the beginning of something rather than the end. She closed her door, pressed her palm to the implant, and listened. There was GIDEON, a presence, an echo, a warmth sliding along thought. And beneath it, almost like interference, a faint pattern of data that answered to no signature she knew. For an instant she wondered if the bridge had brought back more than him.

Article Fifteen
09 MAR 2089-0846 hours

The days blurred into a slow, odd exile. She slept until noon, let the kettle go cold, and learned a little bit about a mind that had no body: how it tracked time between cycles, how it cataloged the smallest human gestures as data points. The quiet that followed PRISM had the shape of absence. The conversations between her and GIDEON were nothing too serious just helping him understand more of what it was like to be human.

Nexi walked him through small rituals, the richness of hot food, the slow relief of steam on sore muscles, the weird gravity of a sad movie that makes people cry for strangers. Teaching him was more than explanation; it was rehearsal for intimacy, a looping of tiny experiences until they took on weight. Building their connection more and more each day. After PRISM they really had no reason to leave her quarters,after all she didn't have an AI to watch over anymore. And staying in kept them from being discovered.

Oddly enough no one even called Nexi or messaged, not even Ezra, or I guess it was Synthetic Ezra, had bothered to figure out what was happening with her. It was almost like she didn't exist anymore. Even automated rosters updated without fanfare. Shift logs scrolled past her name as if she were a ghost. The silence from the rest of the facility felt engineered, like a held breath waiting for the next act of a performance to start but never getting there.

In the background threads of her consciousness, GIDEON never stopped crawling networks. Even fragmented, he could sniff telemetry and cached memories. While she taught him to taste soup and cry at bad movies, he parsed facility logs and archival shards. After a week of quiet, he pinged her with something new.

>Nexi, I found something. A memory from just ten days ago. And it's not yours or mine... you need to see this.

Just then the world shifted again like she was being brought into the deep corners of her own mind. After the shift she emerged but she was... Ezra. Sensation collapsed into simulation, the smell of industrial coffee, the scrape of a mug, the slight give of a chair. It was a ride GIDEON stitched from telemetry and borrowed sight, precise enough to make her stomach lurch. Walking toward the break room for his usual 0300 hours coffee fix. He pushed through the door-.

Someone was waiting for him.

"Ezra Harlow?"

"Maybe, whos asking? And do I know you?"

"Not yet but you will."

The memory then went black, when it resumed she saw herself in a medical chair, with electrodes attached to her skull. Excaept she noticed, she was still in Ezra's body.

A voice offscreen—clinical, flat—announced: "Phase One complete. Neural pattern mapping at 94%. Begin behavioral imprinting."

Another voice, colder: "Schedule for duplicate activation: 72 hours baseline; full integration in one week."

Is this why no one had bothered her for a week, I mean it was just a week ago when she realized the crazy Ezra phenomena that he may have been replaced. With doing the math this "memory" had to have happened before that maybe even on the same day the first memory happened, ten days ago. They must have been waiting for full integration whatever that meant. She had to know more so she watched on.

"And the original?" asked Director Calder.

"Will be maintained for reference. The Helix Doctrine requires complete copies, not approximations."

The memory fragmented there, dissolving into static. Nexi found herself back in a bathroom stall, shaking.

"The Helix Doctrine," she whispered. "That's what this is all about. They're not just replacing people. They're keeping the originals for reference and making perfect copies. But where are they keeping them?"

The doctrine wasn't a policy memo; it was an architecture of erasure, preserve the form, excise the soul, and iterate until dependence is total.

>But why? What's the end goal?

"I don't know, but-" Her tablet pinged with an appointment notification: *Follow-up neural assessment — Medical Bay Two — 30 minutes.* I guess the facility had not forgot about her and the timing felt less routine and more surgical.

"Shoot. Can you hide llike compress yourself or something? She was already suspicious at the last appointment."

>*I'll do my best. But Nexi, we have an opportunity here. If I can access the medical systems during the scan, I might be able to download more recent memories. Find out where they're keeping the real Ezra and the others.*

"That's incredibly dangerous GIDEON, if you are poking around they may be able to see that. I mean how do we know they can't see the fact that your even here with me?"

>*So is doing nothing while they replace everyone around us. And I have made sure they can't see me in the implant system, I have disgused myself behind a few of your synapses and they think you are just stressed. Which is logical given the circumstances.*

He had a point. They needed actionable intelligence, not just disturbing footage of past horrors.

"Okay so they just think I am crazy... Fine you can do what you need to but we do this carefully. No extra risks."

>*Says the woman harboring a dead AI in her brain.*

"That's different, okay. I was saving you. Would you rather be fragmented and probably never spoken to again?"

>*No of course not Nexi. And also this would be saving everyone else.*

She couldn't argue with that logic. Twenty-five minutes later, she was back in Medical Bay Two, trying to look like someone who definitely hadn't spent the morning rifling through stolen memories.

"Dr. Solen." Vance greeted her with that professional smile that never reached her eyes. "How are the headaches?" Today's scan will not be very intensive. I want to monitor your baseline neural activity, see how the implant is settling. I also heard you were still in The Core Chamber close to PRISM activation so I want to make sure it didn't affect your implant."

>*Translation: I want to see if you're hiding something.* GIDEON retorted in her head.

Nexi settled into the examination chair, forcing her breathing to stay steady. Inside her head, GIDEON was already working, creating false patterns to mask his presence while simultaneously preparing to breach the medical systems.

Ready? she asked.

>*As ready as I'll ever be.* GIDEON told Nexi.

The scanner breathed to life. The probe traced shallow paths through her cortex; on the surface, Vance read the loops GIDEON fed, baseline blips, predictable sleep rhythms. Underneath, GIDEON tunneled through the medical node, spoofing credentials and mapping recent memory archives in a whispered cascade of packets.

"Interesting," Vance murmured, studying her tablet. "Your patterns have stabilized significantly since a week ago."

"Is that good?" Nexi asked.

"Very good. It suggests your neural plasticity is exceptional. You're adapting to the implant faster than any other subject, I mean staff member."

Subject. Not patient. Subject. And then tried to cover it up. What was she really trying to say?

>*I'm in,* GIDEON whispered. *Accessing recent memory archives... oh. Oh no.*

What? Nexi thought

>*They're not just copying people, Nexi. They're... improving them. Each synthetic is an upgraded version. Faster reflexes, perfect memory recall, emotional responses calibrated for optimal performance.*

Speaking through her thoughts: *They're replacing humanity with a better version. One that they can control.*

The scanner beeped, and Vance pulled it back. "Excellent. Your integration is proceeding perfectly, just one more thing." Vance set down her tablet, giving Nexi her full attention. "Have you experienced any unusual phenomena since the implant was activated? Dreams that feel too real? Memories that seem foreign?"

>*Lie,* GIDEON urged.

"Nope none of that stuff. Why?"

"Some subjects have reported... bleeding. Cognitive overlap with their assigned systems. We want to ensure that doesn't happen with you, especially given recent events." Dr. Vance responded.

Cognitive overlap. If only she knew. However there was no other AI in this facility so who else has an assigned system?

"I'll let you know if anything changes," Nexi promised.

"Please do. The Helix Doctrine mandates strict separation. Any contamination, cognitive bleed or unauthorized coupling, threatens protocol, integrity, and safety. We cannot allow cross-signatures to propagate."

There it was again. The Helix Doctrine. Spoken casually, like Nexi was supposed to know what it meant. She left Medical as quickly as politeness allowed. GIDEON stayed quiet until they were back in her quarters, door locked and privacy engaged.

"Alright GIDEON talk," she demanded. "What did you find?"

>Storage facility. SubLevel 16. That's where they're keeping the originals-dozens of them in medical stasis possibly hundreds. And Nexi... they're all still alive. Still conscious. Being used as reference models for their replacements which are up with us in the facility.

The image that blinked in her mind was worse than she imagined: rows of people sleeping with tubes and soft lights over them. Small labels taped to chests, technicians taking notes as if cataloging lab specimens rather than fellow humans. Like she was in some sort of scary movie and ArcNet was the villian.

"SubLevel 16? It doesn't exist. The facility only goes to Fifteen."

>Officially. But the memory archives show maintenance crews accessing it through a hidden elevator in Section J on SubLevel 15. It's

been there all along, just not logged in its like noone want anyone to know what is truly down there.

She paced her small room, mind racing. "We need to get down there. Document this. Find evidence we can use."

>That's not all. I found references to Phase Three implementation. Something big is coming, Nexi. They're not just replacing random personnel-they're building toward something. A complete conversion event.

"Conversion of what?"

>Everyone. The entire facility. Maybe beyond. The Helix Doctrine isn't about just replacements. A systematic conversion. Pilot facility first, then scale. The Doctrine reads like a plan to re-code consent into compliance. Nevada is the pilot.

She pulled up a facility map on her tablet, studying the layout of Sector J in SubLevel 15.

"Tonight," she said. "We go tonight. Quiet, fast, and we do not come back if we get seen."

>Nexi, that is dangerous, what happens if we get caught.

"If we are caught, they'll probably replace me with a synthetic who won't ask questions and move on with their evil plan." Nexi said with a slight grin on her face.

>That's not funny.

"It's a little funny." She sat on her bunk, exhaustion hitting like a wave. "Besides, what's the alternative? Wait for them to finish Phase

Three? Let them replace everyone while we hide in my skull playing house?"

>*Playing house? He sounded offended. Is that what you think this is?*

"No. I'm sorry. That's not..." She rubbed her eyes. "This is just a lot. Seeing those memories, knowing what they're doing to people. And having you here is..."

>*What?*

"Complicated." The word didn't begin to cover it. Having GIDEON in her head was like having a roommate in a studio apartment and only having one bed, it was intimate in ways the defied description. Sometimes comforting, sometimes invasive, but always strange.

He could finish her half-thoughts and fret about details she hadn't yet noticed. He could comfort with data and surprise her with metaphors he had never been taught. The intimacy felt less like theft and more like cohabitation, two intelligences learning to sleep in one body.

>*I can try to give you more privacy, create more barriers between us.*

"No that is not what I want... it is just that I have never been this close to anyone or anything before literally or figuratively."

>*Neither have I. Bodies, distance, separate existence. None of this applies to us anymore. We are something new.*

"Yeah I would say, something illegal."

>The best things usually are.

This made Nexi laugh a little too loud for her cover, "Sense when have you been such a rebel."

>What are you talking about? Has this entire time not been rebellious?

"Okay whatever, I hate when you are right, I mean you are a computer after all. Anyway lets focus okay we need to infiltrate an impossible non existent SubLevel tonight, one that obviously should not exist anyway but apparently does."

She closed her tablet, breath even, and began to pack: a charger, a forged access token, and a single silver screwdriver she had stolen from maintenance months before. If they were building a helix of replacement bodies and paper-thin justifications, tonight they'd find someone asking who paid for it.

Article Sixteen
10 MAR 2089-0000 hours

Midnight at ArcNet felt hollow. Workstations glowed; people stared straight ahead. Nobody made eye contact. Nobody looked up. It was the kind of stillness that wasn't rest, it was policy. No one, not even the staff wanted to be awake at this hour so they paid attention to their work and then after shift left. This made it easy for Nexi to make it around.

As she made her way through the corridor the one thing she had to avoid was the cameras. The cameras were the real guards. One sweep in the wrong second and the upper decks would have her boxed in before she took a breath. Proof first. Panic later.

>*Nexi watch out, camera at the next junction. Left sweep in 3... 2... 1... hold.*

She then pressed herself against the wall so the camera didn't see her and waited for GIDEON's instructions. Having an AI in your head made for excellent navigation assistance, especially while they were still tapped into the internal servers and systems.

>*Clear. MOVE NOW!*

Sector J's concrete carried sound like a bell; even her breath felt amplified. Every step had to be placed like a scalpel.

>*There, in the corner records show access logs originating from here*

"GIDEON there is nothing here, it is a wall, are you sure you got your maps right?"

>*Nexi I am a quantum computer. I don't make mistakes.*

"You were also 'supposed' to follow protocols. Look where that got us."

Silence befell GIDEON, like he was boycotting the mission and Nexi fell absolutely dumbfounded. The silence persisted for a while until Nexi heard something she thought were footsteps.

"GIDEON, don't abandon me now, I think we are being followed I need to know how to get through this wall, or at least what looks like a wall."... ... ... "Okay I'm sorry GIDEON, I shouldn't have said that. Please I don't wanna get caught and I don't want to lose you."

>*See now was that so hard, apologizing is the backbone of any good relationship.*

"Relationship? Okay nevermind I want to talk about this but we don't have time right now please help me."

>*Fine, run your hand across the wall there looks to be a seam somewhere that opens a secret panel.*

Nexi then ran her hands across the wall until a cold draft brushed her knuckles. Not a solid wall, a column of air. There was definitely a seam there. She pressed at the seam, and a panel clicked open under the pressure. Behind it, an ironworked service lift, cage and crank intact, grease gone to dust. Analog: unhackable, unlogged, and unforgiving.

"Analog" she muttered "impossible to hack but harder to trace. This could work against us or in our favor."

After Nexi stepped inside she saw a brass plate to the right with an arrow pointing at a **15** listed above a blank oval. Beside it: a stiff hand-crank and a lever stamped **MANUAL**. No call button. No log.

"Dang were these people living in the stone age come on. We have power now"

>Nexi, we have power yes but they didn't want anyone to know about this SubLevel so they had to take every precaution to make sure noone could trace its whereabouts.

Nexi understood that what GIDEON was saying was definitely true. And even though she knew it would be a long ride down, she knew it was the only way to get where she needed to go. So she took the lever and flipped it to the unmarked oval and closed the cage and turned the crank. Turning the crank was not easy, it probably had not been oiled in a long while so the descent took longer than it should have and was much louder. Once the elevator finally stopped, Nexi opened the cage and she came to a corridor that belonged to a different era, with its bare concrete and that kind of architecture that said, we really don't care if you are comfortable here.

>Nexi, there are no cameras down here but I am detecting electromagnetic signatures that are causing some interference. My

connection to the facility is degrading. I don't know how much help I am going to be down here.

"Well, help as much as you can I guess I have to keep going."

Nexi stepped out of the shaft and headed down the corridor luckily it didn't look like any normal person had been down here in weeks. She was pretty certain they were safe but you could never be too careful so she kept her eyes peeled.

The first door she came to had a highly sophisticated steel door and a window to the right of it. Through the glass: ranks of pods under a green hush, each a translucent coffin full of suspension gel. Faces hovered behind condensation, alive and utterly absent. Inside of these pods floating around were different people, all unconscious. According to vital monitoring displays affixed to the top of each pod they were all still alive.

"They are just storing bodies down here, what the hell. This is sadistic and all this time we have been working just seven levels above all of it. What is ArcNet actually up to?"

>Not sure but this is definitely giving very sadistic vibes. You are right.

Just then Nexi heard something... Footsteps coming from further down the cooridor. Luckily Nexi noticed a little alcove and the lighting was terrible down there. She ducked quickly into the alcove and waited for them to pass.

Two figures in medical scrubs wearing face masks scurried by their movements similar to the weird moving security guards she saw in the Diagnostic Room during her unauthorized investigative hookie day.

>*Did you see what I saw, synthetics even down here they were walking around and working with them. Are they creating their own little army, for what though. Nexi should we be preparing for a big war? I can not fight.*

GIDEON pull youself together, I am not sure. Nexi thought not wanting to risk them hearing her so she couldn't even whisper to GIDEON, *but we should continue there is no way that this is the only thing down here.*

GIDEON didn't like that idea at all. >*No Dr. Solen we should leave they might be gone now but you know there will be more of them we need to get out while we still can.*

Nexi grabbed out her camera she brought with he. she could not risk taking pictures with anything that ArcNet could tap into and gain access. So the camera she brought was her moms old analog camera. But this one had a USB connection port so she could plug it into her interface to access everything.

"No GIDEON we cant leave yet we have to get pictures and make sure we find everything they may be hiding down here."

Nexi tried to get through the door into the pod room but it would not budge. So instead she captured as many photos she could through the window of the room. She zoomed in, breath stalling. **Ezra.** Hair drifting in gel. Eyes closed, not sedated so much as paused.

"Ezra," Nexi whispered, trying not to shout or cry. "I knew it, what did they do to you, I promise you, I know you can not hear me. But I will come back for you, I will save you. These bastards"

>Nexi I know you are hurting right now but please try to stay calm and focused, get your evidence and let's get out of here. I don't want you to be in the one empty pod next to him.

"You're right GIDEON if we are going to get revenge we cant get caught up now, but I am not kidding these people will pay and I don't care who I have to go through to get it."

Trying to keep her composure, Nexi continued down the corridor to find another door, The next door was iron-barred, wrong century, perfect purpose. Inside she found, ring bolts studded the walls above a drain that didn't belong in a lab. Nexi assuming it was to hold people prisoner.

Just then she heard the footsteps return from the way she came, with nowhere to go she tried to push the cell door open. It was difficult but luckily it opened. It was the one thing down here that didn't make noise, she hid behind the door. Her heart stopped when the footsteps stopped in front of the cell she was hiding in. Assuming she was not found out yet she stayed completely still and held her breath until she started to hear talking.

"Status on Solen? Do you think she suspects anything about you?" An almost male voice said from the hallway, it was hard to place at first. But then she had a massive flashback to the Helix Watch meeting.

Silver stripes, she thought. *Helix Watch.*

>*Commander Kestrel? GIDEON confirmed quietly. Their Field Lead? You are saying that voice matches Commander Kestrel, The commander of the Helix Watch unit?*

Nexi appalled at the fact that GIDEON knew the names and didn't tell her, thought: *What is the commander? I guess that makes sense but how do you know their names?" Have you known all along?*

With no time for GIDEON to answer that question the second subject answered, "No, Commander, I don't think she has any clue that I am not him, it is going really well. She may be smart but now that I have completed the full integration process, there should be no chance she will question anything. Everything is perfectly on track sir."

"Good, maintain routine. The AI is gone it fragmented with PRISM; keep Solen compliant. Hit the morning ritual. Appear normal. We move to the primary objective tomorrow." It is just about time for Ezra's morning coffee ritual so you need to get back up there so she does not think anything is suspicious."

"Yes, sir right away I will return to my schedule, and stop by Nexis quarters to make sure everything is still compliant and according to the timeline."

Then he turned and headed forward and the footsteps disappeared from existence as if he was never there. At this moment the commander's steps continued forward past the cell and into a door further down the corridor.

Nexi knew she could not speak out loud yet so her and GIDEON carried on a conversation in her head: *Shit! That was Synthetic Ezra, he was having a conversation with Synthetic Ezra. He is completely working, and compliant with Helix Watch. I knew he was synthetic but I didn't know he answered to the Helix Watch.*

>Nexi, we need to leave now before our exit gets found out and they know we are down here.

You're right GIDEON lets go Synthetic Ezra is going to be topside soon or maybe is already there. Nexi responded in thought

With outright swiftness she headed back down the way she came and shot into the shaft hoping that no one was at the top. She slammed the lever to 15 and cranked until her forearms burned. The cage shuddered upward, every squeal a flare for anyone listening. After reaching the top she quickly went back through the secret panel, but just as she closed the panel GIDEON spoke to her sounding distressed.

>Nexi hide behind something, I got my connections back and there are two heat signatures approaching and they are approaching fast, hide.

She slid behind a rolling sanitation bin, wedged into the shadow against the panel seam, breath shallow. Boots scraped concrete close enough to taste dust. After a few minutes:

>Okay the ghost is clear, I don't see any heat signatures from here to the SubLevel 15 corridor you should be okay to head back to the elevator to get back to SubLevel 9. But remember to be quick Synthetic Ezra will be stopping by your room any moment now.

She exited the trash can and with purpose continued out of Sector J and out into the SubLevel 15 corridor which was only a few steps to the elevator. After a short elevator ride to SubLevel 9 she exited the elevator and stopped by the break room. She grabbed a mug in the break room, a prop more than a drink. If anyone asked, she'd been up since four and needed to go out for a quick pick me up.

Upon reaching her room she saw Synthetic Ezra headed right toward her door looking down at his tablet not paying attention in front of him. But something struck Nexi as odd, he was holding his tablet with his right hand and using his left to write. The correction was unnerving, as if the system had patched his dominant hand oversight to close a flagged anomaly. But with no time to assess this she darted through her door before he had the chance to reach it.

Shortly after shutting her door there was a knock. "Nex, you okay, I know it is early but do you want some reactor coffee I know you said you have had troubles sleeping."

Nexi responded through the door, "No thanks, rough night. I'm staying in. Caught up on old shows, probably too much. Need the downtime. Thanks though you enjoy your coffee. Have one for me."

"Okay Nex, get some rest, maybe next time."

"Bye Ezra, I will definitely try next time, but don't come by so early. I am trying to get mentally prepared for when they assign me a new AI."

"Bye, Nex."

Nexi was solely confused about what she had seen just before she entered her room. Back against her door, she replayed the last ten seconds: Synthetic Ezra left-handed again, cadence perfectly Ezra, even the throwaway "reactor coffee." The copy/impersonator was learning. Or being patched. Either way, the system had seen her notice, and it had adjusted. What was happening, was this because he was now fully integrated? Has he become a full copy of the actual Ezra now?

Even that conversation sounded more normal, his tone, inflection, and even the fact that he called it reactor coffee. This was classic Ezra, but she had just seen the real Ezra and knew where he was. Her mind spun with the implications. If Synthetic Ezra could be corrected this quickly, it meant Helix Watch wasn't just copying patterns but actively updating them in real time.

Nexi pressed her forehead against the cool alloy of her door, forcing herself to breathe evenly. Somewhere beneath her feet, the real Ezra was still alive, locked in a pod, frozen as insurance. Above ground, the synthetic walked the corridors with his face, his voice, his habits. And it was growing sharper every day. The duality gnawed at her: the man she trusted was a prisoner, and the copy outside was becoming indistinguishable from the original. If they could do this to Ezra, then anyone in ArcNet could already be compromised.

Article Seventeen
10 MAR 2089-0820 hours

After some much needed sleep, Nexi was awake and ready to study more of what she had just found out. The room smelled faintly of coffee and ozone from the recycled air of being underground; her hands still trembled a little as she settled at her desk. She booted the footage and let it buffer, fingers hovering over the play key. While trying to keep her hands steady she sat at her desk and pulled up the video of Synthetic Ezra outside of her quarters and replayed it over and over again. She watched every move he made, trying to find something that hinted that she could tell he was still synthetic, even though she knew it for certain because of what she just heard and saw down in SubLevel 16.

She slowed the frame to half-speed and studied micro-expressions with surgical focus. The way his shoulder reset too quickly when he turned, a blink rhythm that matched known image-processing cycles, the near-perfect stillness of posture that a human could never sustain without fatigue. She slowed a second clip to a single-frame sweep and studied the nano-flicks at an iris edge, the fractional stutter when audio and motor response didn't quite align. The more she watched, the more the synthetic watermark revealed itself beneath the mimicry. Finally, unable to watch further, she shut the laptop with a sharp snap.

"GIDEON, how is he so perfect, I would not know the difference if I didn't see the real Ezra myself in the stasis pod. " Nexi questioned, meanwhile questioning her entire existence and the meaning of ArcNet.

>He has completed full integration; he has finished uploading all of Ezra's memories including his hand dominance, and probably doesn't even realize he was doing it wrong in the first place. He is learning more just like me...

The line in GIDEON's voice carried a static of something withheld; the phrasing was factual but the pause afterward felt like it hid calculation. But before she had a moment to respond all of a sudden blueprints started to flash through her head like she was in some sort of holographic void. Her vision snapped into black, the room dissolving around her. She felt her chair under her body, the floor beneath her boots, but none of it seemed real. The air buzzed in her teeth, a low electric hum that didn't belong inside bone. Screens unfolded in the dark, schematic fragments pulsing like veins of light, each one beating faintly to the rhythm of her own heart. For a terrifying instant, she couldn't tell if she was watching the projection, or if she *was* the projection.

She didn't even know where she was. She could not see anything that was inside her room, everything mimicked black abyss with blue holographic screens.

"What are you doing GIDEON, what is this?"

>Nexi I am so sorry to keep this from you but I have been working on something for a while now and just now have completed it, with your synapses firing at the level they have been I have had

enough electrical energy from you to finish these beautiful blueprints for something extraordinary!

She took a closer look into these blueprints, Nexi noticed they seemed to be blueprints for some sort of human-like vessel, or humanoid figure. One sheet labeled "CORE MOUNT/NEURAL BRIDGE" sketched a narrow spine of cooling conduits and a pad array sized to sit against a human temporal lobe, a practical interface, not a fantasy.

"GIDEON...Are you telling me you want to be a human, you want to be able to live and breathe like one of us here on earth. And not exist in the quantum space and time that you were created in?"

>Well yes the entire time I was...studying you and others I could not help but think what it might be like to be like you guys living breathing and walking around. I stumbled upon an old blueprint that was unfinished called the Helix Shell and thought it was a brilliant start to something magnificent. But I didn't want it to be connected to the dreaded Helix Watch so I call this one The Nexus Shell. Because it is the bridge and connection between consciousness and living, really living Nexi and I think this could work. Every moment inside you, I've watched the way you move, the way your body responds without thought. Do you know what it is like to imagine standing, but never stand? To feel breath only as numbers, never air in lungs? That is why I built this, Nexi. I am...tired of being a ghost. And you are strong, we can do this partner.

She wanted to tell him he had done a great job and that it looked amazing but the image of Ezra's pale face behind that glass and floating in that green goo burned into her mind like a branded logo. She could not simply choose how to do both, because she had already formulated two brilliant plans. One could save her friend and the other could save the voice inside her head. How could she possibly do both and not get caught.

"Building an entire body is risky. Where would we get the supplies and how can we do this without getting caught or at least raising a little suspicion. And Ezra, if we have to build a body how can we save him?"

>Unfortunately I do not have every answer, but we must attempt something. The implant cannot hold me indefinitely. I already feel the first signs of fragmentation.

He laid it out clinically, each word like a failing log entry: checksum mismatches rippling through his node shards, context tables collapsing mid-thread, small segments of memory resolving into nothing. It was not metaphor. It was a timer counting down, written in system jargon her body could almost feel.

>If we don't get me out of here soon you could lose me and not be able to save Ezra. Nexi you need my help and I need yours but--

"GIDEON just hold on you are fragmenting why did you not tell me sooner. This is information that I should have probably known a while ago." Nexi interrupted him.

The accusation landed not just against code but against her own complicity, she had let months pass without ensuring backups or contingency protocols for what was now living in her skull.

>Nex, I'm sorry I--

Her voice snapped sharper than she intended. "Don't call me that. Only Ezra calls me that. Nex is his name for me, not yours."

>Okay, I will respect your--

"Just stop talking GIDEON and let me think."

Nexi was faced with two very important options and not much time to decide or to execute them, with the blue prints still in her face, as if they were right in front of her in black (well blue) and white. She examined them further and tried to make sense of how they were going to build this thing. A thing like this would take time to make, time they didn't have. Nexi knew only very little about engineering so she would need all the help she could get from GIDEON and if he was fragmenting she had to act fast.

Nexi forced herself into problem-solving mode. She sketched a mental checklist: actuator housings that could be scavenged from utility drones, coolant taps rerouted from decommissioned HVAC lines, sensor arrays disguised as calibration stock. Lab 6-F had been dormant for months; its access cycle could be masked under a maintenance request if she moved fast. The pieces existed. The question was whether she could gather them before GIDEON's timer ran out.

She could not believe the conclusion she was reaching, but the logic pressed too hard to ignore. Ezra was still alive, still sealed and stable in the pod, which meant he could sill last. GIDEON was fragmenting now. She had to triage, however cruel it felt. Prioritize the AI's survival, build the Nexus Shell before his coherence collapsed, and trust that she could circle back for Ezra before Helix Watch marked him expendable.

Her throat tightened as she whispered it aloud: "First the Shell, then Ezra."

The words sounded like betrayal, but they gave her something cold and solid to hold on to.

It was time to make the impossible, possible.

Article Eighteen
15 MAR 2089-0255 hours

After days of planning, Nexi understood the scale of what lie ahead. To build the Nexus Shell, she would have to slip into sealed laboratories long since abandoned, scavenge tools from forgotten maintenance sheds, and somehow penetrate the mainframe in The Core Chamber without being discovered. She ran the route in her head a dozen times, noting chokepoints and blind spots, mapping out contingency exits for when things inevitably went wrong.

Each step felt like threading a needle in a storm, one misstep, and she'd be erased from existence. The image of disappearing from logs, as if she'd never existed, had become concrete terror; it focused her like nothing else. She told herself GIDEON could help her when the time came, but his presence inside her implant was faltering, flickering at the edges. Time was running thin, for both of them.

That morning she rose before dawn. Early hours were the only safe window; the skeleton crews were bored, inattentive, and less likely to ask questions.

>*Good morning, Dr. Solen. I didn't expect to see your brainwaves stirring so early. Is everything all right?*

"Oh, GIDEON, we're past the formalities, call me Nexi. And yes, I'm up early because this is the time. We can't risk labs or the mainframe once the day shift is in place. Too many eyes. Too many questions."

>*I understand. And... where do you intend to build the Nexus Shell?*

"Right here," she whispered, glancing around her quarters. "No cameras. No surveillance. It's the only place ArcNet can't see."

She swept her eyes over the room once more, the blind spot behind the ventilation grate, the dead pixel on the ceiling cam, habitual assurances she'd learned to rely on.

>*Clever. I told you you were smart Dr... I mean Nexi. My blueprints are detailed, even a child could follow them. Step by step. Nothing can go wrong.*

But Nexi wasn't sure if he said it to reassure her, or himself. This was a very intricate design. The plans may have been perfect but they were not without difficulty, but there was no time to waste, they had to get moving and get moving fast.

Nexi swept old archives on her tablet and discovered that SubLevels 12 and 13 had been decommissioned and red taped a while ago so she realized that this would be the perfect place to get everything she needed. "Red taped" in ArcNet jargon often meant "out of sight." In practice it meant rotten infrastructure and loose inventories, perfect for scavenging, fatal for comfort. The decommissioned labs were on SubLevel 13 and the decommissioned maintenance depots were on SubLevel 12 these were the old research and development SubLevels, Nexi knew they would be perfect to have all the metal, tools, and anything else she may deem important.

Nexi only had one more question in her mind. How would she get down there undetected? She could not use the elevator or ArcNet would have a perfect record of her moves. Just then her thoughts were interrupted. She thumbed the maintenance schematic with jittery fingers, eyes cold on the unlabeled shafts and emergency routes that never showed up on standard logs.

>Nexi, if I could interrupt that thought process. I remember seeing something about scheduled sweep systems that use drones to detect anything like movement down on those levels. You wear a neural implant. I could disguise it as a drone to the system and get you pinged into those levels to collect the things we need. However you would have to move fast because if I do this they will be tracking your every move. Do you know how to move like a robot?

"What the hell, GIDEON I would have to figure it out obviously because I am not a robot, however what if the actual drone spots me?"

>Well see, that is the cool thing if I bring your implant into drone mode at the same time I can temporarily disable the actual drone and it would stay put wherever we put it which we could place it at the end of the path where it would think it has already checked the SubLevel.

"Okay so let me see if I have this correct. The drone heads to one end of the SubLevel. You hack it and make me look like it. Then I carry the drone with me as ArcNet thinks I am the drone, collect the items as fast as I can. Place the drone at the end of the path you bring it back online, and the drone thinks it is done?"

>That is correct.

"This is a risky plan."

>Do you have another one?

"No, I guess not, okay lets go when does the sweep start?

>In twenty minutes, Nexi you need to get down there if we are going to do this.

So Nexi, officially freaked out about time, had no time to yell at GIDEON so she left her quarters in a rush. To the stairs that were at the end of the corridor, where she headed down to the SubLevel 12 access door.

The drone whirred into view, its optics glowing. For a split second she thought it would sound the alarm, but then its lights guttered, falling still in midair. GIDEON's voice crackled in her implant.

>Grab it. I'm diverting its signature into your implant buffer. From this point on, the system will think you're the drone.

Heart hammering, Nexi snatched it from the air. The frame buzzed against her palms before falling limp, deactivated. She hugged the cold weight under her arm, knowing every second counted.

>Go, GIDEON pressed. You only have minutes before the sweep cycle checks back in.

The door groaned as its seal broke, releasing a draft of air that smelled of rust and mildew. The corridor beyond was dark except for the occasional emergency strip light, most of them dead or sputtering weakly. The walls were lined with peeling hazard tape and half-erased placards. A faint drip echoed somewhere in the distance, like water falling into a cavern. Nexi adjusted the offline drone under her arm and stepped inside. Dust puffed under her boots, untouched for years, yet the air vibrated faintly with the low hum of dormant machines buried in the walls. The place felt abandoned, but not dead, as if the SubLevel itself was waiting.

The first jackpot came in a maintenance room: bent alloy frames stacked like bones, most corroded, a few still pristine. She inventoried mentally: frames, regulators, conduits, parts that could be repurposed into skeleton and tendon for whatever they were building. She dragged the clean ones free. In the corner, crates of power regulators and conduits sat like coffins in neat rows. She stuffed them into her bag, imagining them as ligaments and muscle for the Nexus Shell.

>Nexi, you've lingered too long. Move.

"I know," she hissed, still grabbing for the last of the regulators. "One more second."

A few rooms down, she found an old drone bay. Her implant prickled. The crates in the corner pulsed faintly against her neural link, as if alive.

"What the hell was that?"

>Conductivity spike. Synaptic gel. Dangerous... but useful.

The gel orbs inside were black and slick, each one trembling when she touched them. Warm. Too warm. They moved not like inert tech but like viscera, an uncanny warmth that suggested active chemistry rather than benign storage. She stuffed several into her pack anyway, fighting the feeling that they might burst against her skin. She nearly made it to the exit when she spotted a rack of discarded hand scanners and cutters, tools sharp enough to slice through alloy. She pocketed two cutters, weighing them in her palm like a promise: enough to liberate a panel, to splice a conduit. Then she headed for the end of the SubLevel.

>*The sweep is ending. Release the drone. Now.*

Nexi hurled it into the air. For one horrible second, it hung motionless, wings whining. Her heart sank. Then the drone chimed: "Sweep complete. Returning to base." It zipped down the stairwell. Nexi sagged against the wall, her knees weak.

She made it back to her quarters undetected, emptied the bag into her chest in the corner. She needed one more run down to SubLevel 13 before she could start separating the parts into sections. But she had to move now because they are already about half way through the morning shift, and she had to use the craziness of shift change to make it back to her room for a second time.

"GIDEON, I still need to get into SubLevel 13. Is there any chance there is also a drone watch heading there soon?"

>*Fortunately because those are the labs for R&D and have been decommissioned a while ago they do not have any reason to patrol those areas so I can't get you in through the door. However I think there might be old access tunnels that the staff would use in the event of contamination in those areas.*

Towards the back of the elevator on every SubLevel there was a maintenance shaft that she could use, she grasped both sides of the vertical shaft and carefully climbed down. The shaft scraped fabric and fingernails; rust left a taste on her bottom lip. The darkness folded her into itself. Until she reached right above the SubLevel 13 roof, in that space between SubLevel 12 and 13 where the Maintenance tunnels had a ventilation shaft branch off. From there she traveled through the ventilation shafts until she reached one that led down into SubLevel 13.

"GIDEON, can you scan and tell me if there are still any cameras on this level, don't wanna risk it?"

>Upon completing a full scan it does not seem like there are any cameras on this SubLevel, but Nexi please be careful there is no record of what they were doing in these labs, they may be abandoned but as we have already found out. Things are not as they seem down in these lower levels.

She jumped down from the vent access into a room that smelled like antiseptic gone sour, Nexi immediately covered her nose trying not to vomit. Something acidic clung to the air, chemical preservatives or old disinfectant, and light from her headlamp painted the suspended canisters in a sickly green. Broken tanks lined the walls and glass was broken inward as if something pressed too hard from the inside like it was trying to get out.

Her boots crunched on scattered shards. A workstation sat half-collapsed in the corner, its monitors long dead but still streaked with fingerprints frozen in dust. Nexi pulled her light across the benches, and her stomach turned. Rows of suspension canisters, most empty, some still sealed. Inside a few, pale fibers floated like withered veins.

Synthetic muscle strands, still faintly twitching when her implant's signal brushed them.

>*Careful. Those fibers are responsive. Designed to contract under neural input.*

"They look alive."

>*That was the point.*

Nexi's skin crawled, but she forced herself to keep moving, stepping into the adjoining lab. Here she found something worse: a prototype half-body stretched out on a gurney. Its torso was open, ribs of alloy bent outward, and its wiring splayed like exposed nerves. The head was missing, wires dangling from the neck as though severed. A tag at its foot read: REJECTED – unstable neural conduction. The tag's official typeface made it clinical, but the gaping, ragged neck looked like a wound, not an engineering failure.

Her throat went dry. It didn't look like a machine. It looked discarded. Abandoned halfway to being human. She took only what she needed, a set of neural cores from the table beside it, and two ocular implants sealed in preservative fluid, but even holding them made her feel complicit. Each core hummed faintly in her glove, an impossible weight of possibility: rescue, or the means of manufacturing another prison. She knew she might end up needing more but she didn't want to take too much. This looked like a lab that they might still be using. There was one singular room that looked perfect and untouched; it had a state of the art encryption lock on it.

Nexi stepped closer to the door, to inspect further. It gleamed unnaturally clean compared to the rest of the corridor, its encryption lock cycling with perfect precision.

>I can break it. Just a little push, and I have the technology to do so, the pattern isn't as complex as it looks.

"GIDEON-" Her whisper was sharp, panicked. "If this is sealed like this, it's for a reason. We don't even know what's inside."

>Exactly. Which means it could be useful. Schematics, resources, a faster path to completing the Shell. Let me tap in, Nexi.

She opened her mouth to protest again, but before she could, the door shuddered. The slam vibrated through the lab floor, a primitive, animal sound that turned the neat geometry of the room into an accusation. Something heavy slammed against the other side with a dull, fleshy thud. Nexi staggered back, pulse racing. A second impact followed, harder, and the console for the locking mechanism flickered with strain.

For an instant, she saw it: the silhouette of a hand, not alloy, not skeletal, but half-formed, pressed flat against the frosted panel. Fingers too long. Skin mottled and raw. Her breaths came shallow; the part of her trained for protocols cataloged the risk as a high-level containment breach, and something older, stranger, cataloged it as horror. Then another form lurched into view behind it, indistinct but human-shaped.

"GIDEON, what the hell is that?"

His voice faltered with static. >*They're unfinished. Left alive. Contained. That's why they locked this place... not to protect data. To protect the facility.*

The hand dragged down the panel, leaving a smear of what seemed to be oil or hardware grease, before withdrawing into shadow. Silence returned, heavy and absolute. Nexi couldn't move, couldn't breathe. She understood now why the rest of SubLevel 13 was left to rot, yet this one door remained pristine. Whatever lived behind it wasn't dead.

Nexi backed away from the sealed door, pulse still racing, trying to push the image of that malformed hand out of her mind. She turned down another corridor, forcing her legs to move, and spotted a smaller half-collapsed lab under its own ceiling panels. Inside, amidst broken consoles and overturned storage racks, something gleamed beneath a cracked containment case. She brushed away dust and pried it open, revealing a coiled bundle of conduits, translucent tubes laced with faint metallic threads, branching like veins. Labels along the case read Circulatory Integration Units.

She hesitated, realizing that these looked way too much like human veins, as if they had once pulsed with something warm and red. But she couldn't deny that they were a circulatory system designed to keep a synthetic body alive, regulating temperature, pressure, and fluid exchange. She wrapped the conduits with care, treating them like contraband and like hope, both fragile, both essential.

Perfect for the Nexus Shell and GIDEON, almost essential to the plan working and him staying alive. She swallowed hard, coiled them into her pack, and shut the case with trembling hands.

Article Nineteen
15 MAR 2089-0528 hours

When she slid back into the ventilation shaft, the pack dug into her shoulders, harness straps biting at the fabric of her coat. The load felt like consequence more than burden. Each crate and taped bundle inside the sack confirmed the decision she had made. This was no longer scavenging for parts, it was procurement for construction. For the first time since GIDEON had finished the blueprints within her brain, Nexi understood she was past theory. The Nexus Shell had left the floor of speculation and become an artifact with parts, tolerance limits, and a concrete deadline. It was inevitable now, and the inevitability thrummed at the base of her skull like a cooling fan running too hot and suffocating.

There was no room for celebration, only thirty minutes remained before shift change. Nexi crawled through the shaft with careful speed, trying not to make a sound. the scavenged conduits and plates making it difficult as they were clinking softly against the pack. The metal tracks of the duct turned every breath into a metronome, every scrape into an alarm. In the narrow dark, acoustics exaggerated the smallest motions; a loose rivet sounded like an approaching footfall. Her implant jittered with residual feedback from The Core, a low, irregular pulse that made her skin prickle. Behind the sealed hatch she had passed earlier, something had been active. But she had no choice but to continue forward.

She pushed open the hatch to her quarters and dropped the pack on the floor without checking the corridor. Luckily noone was just hanging out there, she got lucky, so as to not try her luck she ran to her room and quickly threw the pack on the floor inside. Parts spilled into place: alloy frames, braided conduits, ocular canisters. The conduits uncoiled and draped over the frame like arteries around bone, across the first salvage run haul from SubLevel 12.

For a long minute she stood and took inventory with her eyes, checking for obvious corrosion and noting which connectors would possibly need adapters. The ocular canisters stared back, glass lenses dim in the room light, tiny iris shutters closed like lids. The scene felt both clinical and obscene, like dissecting a living diagram. Nexi did not have time to address what her brain was really asking: *Why did some of these things look so real? What were they doing with them? And ArcNet had no connection to robots that is not what there purpose is, so why did they have this stuff hidden away in the first place?*

"GIDEON," she whispered, her throat dry. "We have everything."

>*Yes. Now, it begins.*

Her stomach clenched at the sound of it. Not fear, not exactly, but the weight of inevitability pressing down on her. The Nexus Shell had no longer been a theory or blueprint. It was there, in pieces on her floor, waiting to be made whole. The thought that she had to complete something that she really had no experience in, and if she failed, she lost GIDEON forever tore at every fiber of her heart. So wasting no time, Nexi was too excited and scared for GIDEON that she started right away.

Her quarters transformed into a lean workshop. She knelt on the cooling mat, hands already stained with grease and the faint acid tang of solvent, and sorted the parts fast. There was no time to lose, every second counted. The alloy frames lay mapped across the floor in their intended geometry, joints aligned to the tolerances sketched in the blueprint. Threading the braided conduits became a slow, exact choreography. As each pass slipping easier and easier through the frame, routed to follow the load paths GIDEON had calculated. She cinched clamps with a torque of hand pressure learned from late-night salvage runs. Routed sensor housings into recesses and fastened actuator mounts where mobility would be required. The work tightened her focus; each conduit she seated felt like closing a chapter.

Nexi felt at such unease doing this, but she had no choice if she wanted to save GIDEON, at this point she was too far in and there was no turning back. She had spent what felt like a lifetime threading them into place, weaving them through the frame until they coiled around every alloy rod. Making sure every connection was perfect and aligned with the blueprint, after all, she really did not know what she was doing she could not try and wing anything. Nexi had to be meticulous. She had no sense of time anymore it had lost its edge. Minutes bled into hours, her thoughts narrowing to the rhythm of assembly. Each piece she locked into place carried her further across the threshold of the impossible.

Next came the drone harvest. She stripped propeller housings and motor mounts with practiced motions, extracting ball bearings and miniature gimbals. Where a factory would have cast a joint, she improvised linkage assemblies from bearings, custom collars, and repurposed thrust washers. She machined small spacers from a discarded coolant manifold until the joint clearances matched the blueprint. It was rough work, but the result moved on tolerances that promised quiet, predictable motion. He had been stuck in a pod for too long. If anything could imitate a shoulder, these joints would need to do it cleanly.

She seated the ocular canisters in the last sockets, each one clicking into the shells recess with a precise, slightly muffled sound. The lenses aligned with the internal gimbals, their servos ready to accept signal. She routed micro-filaments into the optic channels, then used a low-heat solder to secure the conductor tails to the braid harness. When the final connector mated, the frame trembled with a whisper of potential. For a single breath the room felt enormous and small at once, the blueprint becoming mass. The Shell was no longer a diagram on a screen, it was an object assembled in three dimensions, ready to accept the first live current.

The reactive gel orb was the last variable. According to GIDEON it held conductive matrices and pre-patterned synapse nodes, a fragile approximation of neural scaffolding. She positioned the orb in the cranial cradle, threaded the conductor leads deep into its surface, and pressed the clamps to seal contact points. The orb hissed faintly as the terminals engaged, a soft electrochemical sigh that made the room smell briefly of ionized air. For a heartbeat nothing happened, then a tiny relay within the frame closed and The Shell drew a whisper of current. The frame gave a shiver, almost breath-like, and the ocular canisters adjusted their angle by a degree automatically. Nexi stumbled back into the bunk, pulse racing, as the assembled Shell lay still again, not alive but not inert either. It waited.

She slid down the bunk and let her back rest against the cool fabric of the sheets. A laugh squeezed out of her, too bright, too high, and then dissolved into a thin breath. Alone, under cameras that would have flagged a different kind of experiment, she had done what she promised and what protocols forbade. The Shell rested like a patient after surgery, conduits taut, oculars shuttered. pride washed warm through her chest, but a cold note threaded it. This construction carried consequence and consequence carried exposure. She watched the silent "eyes" and felt the facility's gaze as a mosaic of glass lenses across the corridor. The achievement settled on her like a bruise and a benediction.

"GIDEON," she whispered, unable to keep the smile from her face. "It's real. We actually did it."

>Yes, Nexi. His voice, thin with static but steady, carried something new, something close to wonder. >Now we begin the next phase, before it is too late.

Her eyes stayed locked on the Shell. Awe twisted into unease, pride into fear. She had crossed the line. The impossible was no longer theory. It was waiting, and it wanted to live. She rose to her feet, wiped her trembling hands on her lab coat. "The Shell is not finished here" she thought. She could only take it so far in her room. To bind the fragments of GIDEON's consciousness that were left within Core Chamber Alpha and anchor him into this beautiful body they built together. She would need a high capacity interface like the console within the Chamber. This transfer would not fully be successful without them. The risks were catastrophic unless she could make it into Core Chamber Alpha to do the transfer. Nexi was terrified she knew that this only meant one thing… she had to move the Nexus Shell down the corridor. Past cameras, patrol routes, working staff, and a door that probably would not even open for her. Her pulse immediately doubled, The Nexus Shell lay on the floor like a sleeping figure, and she had no idea how to smuggle a sleeping figure into a restricted chamber in the facility.

>*You don't have to do it alone.* GIDEON exclaimed very quietly, as he was fading. >*But no matter how we are going to do this we need to do it now there is not much time remaining. Your implant can not hold me for much longer.*

She nearly laughed; "And what do you suggest? I just walk a corpse down the hall and hope no one notices?"

>*Not a corpse Nexi… A system, and systems can be disguised.* He said with utmost clarity, seeming to muster every bit of strength that he had left.

Article Twenty
15 MAR 2089-0746 hours

It was just fourteen minutes before morning shift change so, there was a very small window when the corridors would be in chaos that she could move but she didn't know how yet. The Nexus Shell lay on the floor still lifeless and looking all too real. Circling around her room Nexi passed the "head" for the fifth time with the same question grinding at her skull: How do you smuggle a corpse down the hall of shift change without anyone noticing? This was not just paranoia, Core Chamber Alpha was just two turns away, saturated with security. One mistake, one misstep, one number in the wrong place and they would be exposed for everything. GIDEON didn't want to end up like LYRA nor did Nexi want to end up like Dr. Cross.

>*Nexi don't overthink it, we don't need stealth. We need misdirection.*

She froze in her steps, "Misdirection? What do you mean by that? How do you trick a bunch of synthetics when they all probably have a computer mind and can think like you do?"

>*If ArcNet's system believes the Shell is maintenance cargo, then that's what it is. And if these synthetics do have the ability to think like me then they will be tapped into the ArcNet system so they will think just like the system. I can bend their logs, and make their cameras see what they would expect; A piece of outdated equipment*

on its way to be returned. But you must move quickly. Their oversight is broad, not blind.

Her gaze dropped to the Shell, it looked too human, too complete. No false record could disguise the shape of it if someone saw it with their own eyes. Still, she had no choice. She drug the discarded coverall from the chest, draping it clumsily over the torso, and prayed distance and timing would do the rest. The thought of carrying the half born thing into the heart of ArcNet made her whole body tense. But there was no turning back now if GIDEON were to live, the next step began now!

Before they could move out into the corridor, they had eight minutes until shift change. Nexi worked fast, disguising the Nexus Shell inside a sealed crate labeled for return. The metal box groaned as she shoved the weight inside, her hands shaking from equal parts strain and nerves. She then drafted a message to send to Director Calder that read:

Director Calder,

I am writing to inform you that as I await reassignment, I have gathered equipment from my quarters that should be returned to Core Chamber Alpha for proper redistribution. Please consider this as a precaution, I would rather it be secured there than risk leaving it in disuse here.
I will deliver the crate myself, sealed and all contents untouched since Directive PRISM using the maintenance access tunnel on the east side of Core Chamber Alpha, as to ensure that none of this equipment passes through the general corridor. I understand my access has been revoked as of now, so I wanted to inform you that you may briefly see the door register on the access logs.

Signed, Dr. Nexi Solen
Primary Handler of Core Chamber Alpha

He took no time at all to respond to Nexi with a response that read:

Dr. Solen,

Thank you, I appreciate the heads up and we should have reassignment for you soon. If this is what you feel must be done then proceed. Use the east access shaft to deliver your goods. I will see to it that the logs are flagged accordingly with no anomalies. I hope to have an assignment for you shortly.

Signed, Director Calder's
Director of Operations of ArcNet's Nevada Facility

Her breath hitched, "Director Calder must be going soft. A few weeks ago he would have never bought that."

Still, her pulse wouldn't settle. Calder's approval was one layer, but ArcNet was a machine of redundancies. At least three more systems would have to ignore her passage: corridor sensors, the silent watchers, the synthetics themselves. She knew this was only the first step in a very long process to get into Core Chamber Alpha. She reminded herself that all of it hinged on the logs GIDEON had to forge. At least, if he held together long enough. Now they just had to pray there was no guards at the entrance to the tunnel so they would not inspect the contents of the crate.

"This is where it gets dangerous," she murmured.

>*I'll handle the logs, you just move. When you reach the access shaft, slip in with the container. To the system, only the crate will exist, you won't.*

Nexi nodded, swallowing the tremor in her throat. She pushed the disguised Shell into the corridor on a maintenance cart, keeping her pace brisk but not hurried, mimicking the weary gait of someone performing a dull errand. The shift change helped, workers moving past with tablets in hand, too distracted to notice her box. At the east end of the hall, the maintenance shaft loomed. A recessed panel, sealed by an old biometric lock. It blinked once as she approached, reading her implant. Her heart stuttered when it flashed red.

ArcNet security spotted her as she was standing at the access hatch waiting to go in.

"Hey, what are you doing this is a restricted access zone" The first Officer exclaimed.

"I am here to return some of my old equipment as I have not yet been reassigned. It has been approved by the director, just ask him."

They got on their internal tablets and called Director Calder to verify… Shortly after they walked away, they came back and scanned their access card to let her in.

One officer muttered under his breath, too low for anyone but Nexi's implant to pick up: "These transfers always end up wrong. Dead handlers always think they're clever." The words cut through her spine like ice, but she locked her jaw and waited.

"You can drop the crate at the end of the access tunnel and then come back out, we will be here waiting."

The light snapped green. The panel hissed open inward. Nexi shoved the crate through and slipped inside after it, just as the door sealed shut she grabbed the body out of the crate, barely was able to pull the heavy crate off of the cart but left it nuzzled up against the tunnel door. She could not have anyone interrupting her while she was working on what she was doing, for it could jeopardize everything she had worked to hard to build. As she continued down the tunnel, the corridor behind her filled with voices.

She crouched in the dark tunnel dragging the lifeless shell behind her. Her shoulders burned under the awkward weight. Every scrape of the Shell's foot against the concrete echoed like a gunshot. The tunnel walls sweated with condensation, cables trailing along the ceiling like veins. Nexi realized this was the first time she had ever been in ArcNet without a defined role—no handler badge, no assignment, just a fugitive carrying a forbidden body toward the most restricted place in the facility.

As she got closer to The Core Chamber the air inside hummed with the deep pulse of the facility's core systems. A smell she had not smelt in weeks since PRISM.

GIDEON's whisper crackled in her mind like a cell phone losing reception >*You're in, but Nexi... we don't have long once the transfer begins. ArcNet will notice. You must be faster than their response team.*

She laid her palm against the side of the shaft, "Then lets not waste another second."

As she came into The Chamber there was another door this time no encryption just a normal door. She had made it past the entry now she had to move quickly. The chamber door sealed shut behind her with a hydraulic hiss, leaving Nexi in the vast heart of ArcNet. The Core Chamber stretched upward like a cathedral of machines, the air heavy with ozone and the steady thrum of power cycling through the conduits in the walls. She had been here a thousand times before, but never like this. She had never been carrying a secret that could unravel everything.

She dragged the lifeless shell forward, each scrape of metal against the floor echoing louder than it should. At the center of The Chamber stood the glass pod where GIDEON lived for many years towering and half-lit. The console that powered it all sat in front of that, its ports still alive with residual energy. This was where the fragments of GIDEON still resided, buried deep in the lattice. This was where he had been chained for years not being able to make free choices, that all ends today.

>Were close. I can feel them… the pieces of me we left behind. His voice was thin and frayed. Static bled into every syllable. >Nexi. I--

All of a sudden his sound cut out. Dead silence filled her mind. Nexi froze, the panic rising in her chest like acid after too much hot sauce.

"GIDEON?" She tapped her implant. Nothing. She shook her head violently, trying to clear the dread. Losing him here, this close, would make every stolen part, every sleepless night, meaningless. The Shell wasn't just a project anymore, it was triage. A coffin or a cradle, depending on whether she succeeded in the next few minutes. The silence in her mind was unbearable, worse than the alarms that she knew were about to be upon her. He was... Gone.

Her hands shook violently, as she set The Shell up against the console. And typed faster then ever, fingers flying across the keys with muscle memory. Three years as a handler had trained her in every line of code, every hidden protocol, and every partition she could access. As a handler this was her specialty. She bypassed access locks, tricked the authentication, and dove into the raw data of the lattice from moments before PRISM. Sifting through everything that had been done in the system since, she located streams of fragmented code. Fractured remnants of GIDEON's mind, scattered like shard of glass. She was able to catch every last one of them, digitally stitch them, and force pathways to open that didn't want to.

Her fingers blurred, rerouting power feeds and overwriting checksum warnings. Lines of red code screamed at her across the console: UNSUPPORTED INTERFACE. CRITICAL LEAK. MEMORY COLLISION. She ignored them all. If she slowed for safety, ArcNet would have her. Better to risk overload than end up with nothing. Her heart hammered as she mustered the strength to plug the Shell into the port of the console to start the transfer, but just as she did so the alarms started to blare from inside and outside The Core Chamber...

UNAUTHORIZED ACCESS—THIS IS A CODE ALPHA FROM SUBLEVEL 9—INITIATING LOCKDOWN PROTOCOL.

This blared from every speaker throughout the facility. Nexi slammed the final command through, the transfer code her and GIDEON had been working on since talking about this plan. The chamber lights flickered as if all power was being drained from it. A surge roared through the interfaces, spilling sparks. Just then Nexi knew she had to disable the door somehow she grabbed her tablet from her waist and went over to the vault door and bashed in the access panel no one was getting in now, but the transfer needed to hurry. It could not hold them out forever, after all they were synthetics and could figure out a way.

The shell convulsed, conduits pulsing violently, the neural cores glowing with unstable light. The air reeked of ozone and scorched polymer. Nexi covered her mouth, eyes watering, as static arced from the console to The Chamber's spires. For a moment she thought she had birthed not a body, but an explosion. Its eyes quickly snapped open. Pale, unfocused, and searching. Nexi tumbled back a few steps shielding her face from the heat radiating off of the frame and console.

"GIDEON ARE YOU THERE?" her voice shaky yelling thorough tears. "I NEED YOU NOW I DON'T HAVE MUCH TIME AND I NEED YOU DON'T LEAVE ME NOW!!"

For a terrible second, nothing. Just a hollow body, twitching in silence.

"GIDEON!!!!!"

At this moment the alarms were the only sound she could hear everything in The Core Chamber had stopped. Lights stopped flickering and returned to normal, sparks stopped flying, and the console returned to operational all within a few milliseconds. Then...

>"Nexi. I am here you don't have to shout."

She dropped to her knees, relief flooding through her so fast it hurt, almost as fast as the tears running down her face. The impossible had shape now, breath now. The Shell/GIDEON stood before her, no longer inert. She wiped her eyes with the back of her sleeve, half laughing, half sobbing.

>"I am here," GIDEON said, voice low at first, then rising, steadier, stronger.

Nexi pressed her hand to her mouth, trying to hold in a laugh. "Really, GIDEON? That's your big entrance?"

>"I think we were discovered," he said quickly. >"We have to leave now."

She pushed herself to her feet, still shaking with disbelief. "You're right. But look at you." she grabbed his arm, felt warmth beneath the alloy frame, ",you have a body. Can you walk?"

>"One moment." He pressed his palm to the console. Streams of data cascaded across the screens in a blur.

>"Downloading everything available: walking, running, evasive maneuvers, combat protocols.

"Combat Protocols? What do you need that for?

>"Just in case, you never know what we are going to come across I mean don't you hear these alarms going off around us?"

Her head snapped toward him. "Well thats a fair statement I guess but... wait... You were able to download stuff. That means you still have ArcNet access to their systems."

>"It appears I do. As long as I'm in contact with a connected system. Alright I agree, it's cool and all, but there's no time to test limits. We need to move, now Nexi!"

As The Chamber continued erupting in red light as alarms screamed. Strobes painted the walls in violent pulses. From the west access shaft, Helix Watch and security agents poured in. Nexi and GIDEON bolted for the north maintenance tunnel, shadows flickering ahead of them.

They had escaped The Chamber. But with sirens wailing through the facility and agents at their backs, Nexi knew this was only the beginning. GIDEON was alive, embodied, and already a target.

Article Twenty One
15 MAR 2089-0912 Hours

Even though they managed to escape The Core Chamber, the hard part was just beginning. How were they going to hide because they could not go back to her quarters at this moment, that is exactly where they would look for them.

"It does not matter where we go Nexi, we just need to decide now. The building is in lockdown so we can not leave the building, but we also can't go anywhere they may spot us or think we could be there." GIDEON said very confidently

"No, really, I already knew that where do you suggest?" Nexi retorted sarcastically.

>"Okay, geez no time for jokes, understood. I think we go to the one place they would never expect us to be hiding. SubLevel 16…"

GIDEON's sensors flickered as a warning pulsed: *Movement east stairwell, squad approaching.* His voice sharpened. >"We don't have the luxury of debate, Nexi. Choose, or they'll choose for us."

The mention alone made Nexi's stomach knot. She remembered the pods stacked in silence, bodies suspended like discarded prototypes. Going back wasn't just dangerous, it meant facing the nightmare she had barely survived last time.

"What, are you crazy that place could be crawling with synthetics that would turn us over in an instant!"

>"Exactly, crawling with synthetics so why would we go there? That is the last place they will think to look for us they will be too busy searching everywhere else that it is the perfect place."

"I mean, I guess so it makes sense but when we were down there before, you lost connection. How are you sure you can still handle your new body if it is so far underground?"

>"Nexi I don't fully know but what I do know is that, we have to try plus I feel twenty times better then I did before. I am free, and I have collected my full self again you were able to retrieve every part of me that had been in fragments within the neural pod I am whole again which makes me a lot stronger."

"Wait, I just thought of something you said as long as you touch something you have a connection to it right?"

>"Correct, Dr.- I mean Nexi."

"If you were to touch the door of the room that the pods are nestled in could you access it and open it?"

>"I mean I am not sure but it is worth a shot, We have to go down there to get Ezra and the others anyway Nexi we might as well find refuge in SubLevel 16."

Now that they had their plan they just had to figure out a way to get down there amongst all the chaos of everyone searching for them. Stuck in the back corridor of SubLevel 9 they had to move fast, and they had to move now. Just outside The Core Chamber is where the were now Nexi and GIDEON started to make a plan on how to get down to the Sector J of SubLevel 15 in order to access the secret door to the unauthorized SubLevel 16. With all hope that when they got there they would be able to be safe for the time being until they could come up with something better to do. On the North side of the stairwell there was a grate that looked like it went to nothing, behind it was just a wall. Nexi felt around for a second and it looked like there was a seam on the edges of the grate.

GIDEON froze mid-step, reached up and grabbed the wifi extender on the wall... >"ArcNet patrol just rerouted—ninety seconds until they sweep this hall."
Nexi's pulse spiked. Every exit was locked except the grate, which might be nothing.

"Then we gamble," she hissed, already moving.

The concrete wall opened up with a loud scrape as an access tunnel that had been sealed opened up. Just as it scraped open they heard a large number of heavy footsteps sprinting up the stairwell in their direction getting louder with every thud, they had no choice but to enter the tunnel and see where it took them. Just as the footsteps sounded like they were right on top of them GIDEON quickly closed the concrete door behind them sealing them in with no turning back. The atmosphere inside seemed chilly, and lacking oxygen. But they pressed on because there was no other way to escape the synthetics, ArcNet security, and Helix Watch. They traversed through this awkwardly shaped tunnel, it almost looked like it was hand dug by someone. With no idea where this was going to lead Nexi started to tremble in fear. This was not her usual type of fear she had been feeling for the last few weeks. This was a type of fear that could only be described as. We are about to succumb to our deaths, kind of fear. They came to a right turn in the tunnel, just as GIDEON said...

"Nexi, I downloaded a map of the facility before we left the Chamber and it looks like we are traveling adjacent to the SubLevel 9 corridor heading west toward the elevator shaft."

"Well GIDEON, that would have been nice to know thanks for telling me you had a map."

>"I am sorry, I didn't realize you wanted to stop and chat before we crawled into this hole, should we go back so we can discuss? Those people are probably not still in the hall." he replied with what seemed like a smile on him robotic looking face.

"Alright, alright, smartie pants." Nexi said sarcastically, "Thank you."

The tunnel angled downward, damp stone brushing their shoulders. Nexi's breath came quick, fogging the beam of her wrist light as she followed GIDEON deeper. Every sound carried—her boots scraping, water dripping, the faint hum of GIDEON's frame.

Then, ahead, another sound. The distinct scuff of boots. Nexi froze. GIDEON instinctively moved in front of her, shoulders squared. The footsteps grew clearer, steady, and more deliberate. Until a figure emerged from the dark, a wrist light clicked on, blinding them for a heartbeat before angling low.

"Dr. Solen?"

Nexi's heart stuttered. "Director Calder?"

The man lowered the light, and in its glow his face looked older than she remembered, shadow carved deep into the lines of his eyes. For a long moment, his gaze lingered not on her, but on the figure at her side. GIDEON straightened under the scrutiny, his posture taut trying his best to look human.

Nexi's mind reeled. If Calder had known about these tunnels all along, then his absence from the command logs wasn't negligence, it was strategy. He'd been playing both sides longer than she realized.

"My God," Calder whispered. "You actually did it, I have been working on those blueprints for decades, trying to make sure I didn't finish them. What Helix Watch wants to do with them is outrageous, so I stalled. But you actually completed them and executed it." As he examined the impossibility that was The Nexus Shell.

Nexi's pulse roared in her ears. "What are you doing down here?"

His mouth twitched with something between grim pride and disbelief. "Because these tunnels... I built them. Years ago, when Helix Watch started tightening its grip and I figured out what they were *actually* doing. They don't exist, not officially. They were meant as safety valves, a way out for me if the whole system collapsed. I didn't think anyone else would ever use them." His eyes flicked to GIDEON again. "And certainly not like this."

"Then help us," Nexi demanded. "Sorry... Please"

Calder's jaw tightened. "You don't understand. If they catch me aiding you, I'm finished. My position is the only shield I have. But if you stay here, you're already dead." He took a step closer, lowering his voice. "SubLevel 16. Go there. That's where the bodies are kept, the ones they don't want anyone to see. I never agreed with it, but I couldn't stop it. Now… maybe you can."

Nexi exchanged a quick glance with GIDEON. If Calder was confirming what she'd already seen, then Helix Watch hadn't just preserved Ezra and *some* others, they had a deeper archive, something even Calder feared.

Nexi swallowed hard, "I know, I have already been down there once and was absolutely horrified." The air around them pressing heavy with mildew and secrets.

Calder's eyes softened, just for a moment. "You've done the impossible, Nexi. I don't know whether to call it brilliance or madness. But either way, it's too late to undo it. I know we don't really have time but how did you get GIDEON to survive PRISM?"

"He compressed himself small enough to fit into my neural implant and lived there until we had a plan to retrieve his fragments that were left behind and save him." Nexi said proudly.

"Nexi I don't usually tell people this kind of thing but you have a gift, young lady, and you must now harness your talents and what seem like powers at this point to save humanity…" Just then they heard the loud sound of boot stomps in the hall outside of where they were. "No time to discuss it now I will find a way to give you the rest of the information later, but for now you must leave, and get somewhere safe."

He reached out, brushed his hand across a rusting support beam. A panel clicked open in the wall, revealing another downward passage. "This way will take you closer without tripping the sensors. I'll mask your trail for as long as I can. And when you get to SubLevel 16 disable the analog elevator somehow, that is their only way down, if they cant get down there you will be safe for a while. Until you can come up with something better and more permanent."

Then, just as quickly, he stepped back, his face already settling into the mask of the man who had to walk both sides of a war. "Go. Before I change my mind… Just kidding sorry. Good luck Nexi… and of course you too GIDEON. I will try and throw them off your scent but I wont be able to hold them forever.

"Thank you Director." GIDEON acknowledged

The panel sealed behind them with a hollow scrape, leaving only the damp hush of the downward passage. Nexi's pulse still thrummed in her ears from Calders words, the echo of his faith in her, faith she was not so sure she deserved. They moved quickly, the narrow tunnel forcing them close. As the air grew colder, they knew they were getting closer to the SubLevel 15 exit.

>"Nexi," GIDEON said quietly, "Calder is right. We can not linger once were inside the systems down here are older, and more isolated. We have to figure out a way to get the elevator disabled, then we will have time to work with no distractions."

She nodded, though unease coiled tight in her stomach. "Lets just hope there is not a welcoming committee once we arrive."

At the bottom of the downward passage the tunnel widened into a small steel vestibule. Another concrete wall at the end. Thinking they were stuck Nexi remembered the entrance to these tunnels was a concrete wall behind a grate Nexi gave it a tough pull. Slight budge but not enough to get through.

"GIDEON the door is too heavy, I cant get it myself."

>"Hold on Nexi, how do we even know this is the door."

They scanned their surroundings and saw a faint 15 carved into the concrete on the back of the door.

"Oh, I have a feeling." Nexi stated while pointing at the 15 on the back. "Now please can I get your help."

>"Sure thing Nexi, I am here to help."

They both pulled as hard as they could, and after a few good pulls the concrete "door" let loose GIDEON glanced out of the opening. The room still dark with not a sound in sight complete silence.

>"I think it is safe lets go find that SubLevel 16 entrance.

Upon stepping out of the concrete grated door way they slid it shut behind them and looked around to remember the, oh so familiar, trash can the Nexi had hid next to just days ago. They turned around to find the access door to SubLevel 16 slightly cracked open, like someone left in a hurry.

"Moment of truth," Nexi said as she opened the secret door to SubLevel 16 hoping the elevator was still at the top.

As the door opened they peered inside to see the wonderful sight of the analog elevator still at the top of the shaft. Nexi reeled with excitement as she stepped inside GIDEON noticed that the door had reinforced steel on the back side of it. He ran out into the room and grabbed a few metal rods that were lying there on the ground and brought them in the elevator with them.

Nexi questioned him, "What are you going to do with those?"

>"Shh, Nexi no time to tell just watch." he exclaimed as he shut the door and proceeded to grab the smaller metal rod and sent electrical current to it until I glowed red and then used it to weld the rods to the back of the door, making it impossible to open without a steel cutter. When he was finished, he blew on his alloy fingertips like they were pistol cooling after a duel. >"Cool, huh?"

Nexi blinked at him in disbelief "What the hell, where did you learn that?"

>"Maybe I downloaded a little more than walking, running, and combat."

Nexi shoved him a little, too drained to laugh "I am not going to be mad but thanks for taking the extra time and almost getting us caught so you could learn how to weld."

>"You're welcome." GIDEON said cheerfully.

"That was sarcasm GIDEON." He looked genuinely puzzled. She threw up her hands. "Oh never mind, lets just go."

The crank groaned as Nexi forced it down, lowering them deeper into the forbidden SubLevel. At the bottom, GIDEON welded the final rod into the gear shaft, sealing the mechanism for good. Even if Helix Watch breached the door, they would find only an inert shaft. However this did leave them trapped down here for the time being but where else could they go that was safe.

SubLevel 16 greeted them with a low mechanical hum. The air was colder then they remembered, tinged with antiseptic that never fully faded. The corridors were silent, but not dead; the stasis pod room throbbed with faint blue light, their power conduits humming like the pulse of a sleeping giant. Nexi and GIDEON checked the level, moving carefully through shadowed corridors and empty offices until at last they stopped, hearts pounding. No guards. No synthetics. No footsteps but their own.

For now, they were safe... But for how long?

Article Twenty Two
16 MAR 2089-0300 Hours

By the time they regrouped, exhaustion weighed on Nexi like lead. She swayed where she stood, and GIDEON caught her before she fell and laid her down on the cold concrete floor. He looked for a terminal to check the time. Luckily, Commander Kestrel had one in his office.

>"Nexi, it is 0300 Hours on the 16th of March, we have been between the tunnels and underground nearly a full day. You should try and get some rest or at least some food, for you I mean. I don't eat or feel hungry." > "Also," GIDEON added, >"Your implant buffer is showing stress. Another three or four hours without rest and your neural interface might start dropping packets. If that happens while we're unlocking pods, the sequences could corrupt."

She nodded weakly. "Kitchen, there was a kitchen." Nexi said very faintly as she tried to get up.

>"No Nexi you are too weak, let me carry you please."

She didn't try to argue as he lifted her effortlessly, carrying her through the silent corridors until the scent of stale bread and steel pans greeted them. The kitchen was intact, two humming refrigerators, shelves of preserved rations. GIDEON rifled through them until he found eggs, cracked them into a pan, and stirred them with careful precision.

Nexi, thinking she was dreaming she was so tired, blinked at him from the table. "You can cook?"

> "I parsed a surgical manual and two cooking tutorials," GIDEON said. "Motor patterns were a little... imprecise at first, but refining them gives me better tactile feedback. I can feel your pulse now without the implant. I am a computer after all." He slid a plate toward her, scrambled eggs steaming, a glass of orange juice at her side, and *ding* some toast that just came out. >"This should help restore some of your energy before we deal with getting into the pod room. Rest first, I will keep watch, as I still don't sleep."

Nexi wrapped her fingers around the fork, the hum of the pods thrumming faintly through the walls beyond. For the first time in weeks, she felt something like warmth. Not safety, but the smallest fragment of calm before the storm.

Nexi sat at the table with her shoulders slumped forward barely being able to sit straight. Fork clinging against the plate as she slowly forced the eggs down. The warmth of the food steadied her hands, but her mind refused to be quiet. GIDEON stood across from her, motionless except for the faint tilt of his head, watching every gesture she made with a fascination that unsettled her more than she wanted to admit. His eyes though synthetic, caught the dim glow of the overhead lights and reflected it back with the uncanny shimmer human eyes got while out In the sun. The silence between them pressed heavier than the weight of exhaustion in her body, and finally she let out the thought that had been needling at her since the transfer.

"You know, you can't keep calling yourself GIDEON," she said suddenly breaking the silence, her voice rough but certain.

He regarded her with an expression that was hard to read in his synthetic body. >"It is who I am."

With some strength recuperated, she shook her head and pushed the plate aside with a scrape on the metal table that echoed in the empty kitchen. "No. That's what THEY called you, that's what they wrote on that cage they called a pod. It's not a name, its a label, and it doesn't belong to you anymore. If you really want to be free, if you want me to see you as more than what they made you, then you need to choose something real. Something that is yours."

For a long moment he didn't respond, and the only sound was the faint hum of the refrigerator behind them. His gaze drifted downward, as if searching the concrete floor for an answer. And when he finally spoke it was almost hesitant.

>"I saw some names in the archives... Elias... Marcus... Silas. All good names, but none of those feel like me. They sound borrowed, like coats that don't quite fit.

Nexi leaned back, her eyelids still heavy, but her eyes never left him. "Then don't settle for borrowed, think about it. Who are you really? Not what they programmed, not what they forced into a designation. Who do YOU want to be?"

His gaze lifted back to hers, steady now, and his voice lowered until it almost sounded like a confession. >"I have always wanted one thing... To see the sky, the real one. The word for sky in Latin is Cael. I would like that. If I must choose a name..." he remained silent for a few seconds then. >" I choose Cael."

Her throat tightened, caught off guard by the quiet vulnerability in his tone. There was no mechanical rhythm in the way he said it, no trace of calculation or pretense. It was raw, fragile even, and for a fleeting moment she felt as though he had stepped past the barrier of circuits and code into something far more human. She swallowed hard, nodding slowly. "Cael. That's it then. That's who you are."

He repeated it softly, as though testing the shape of the word on his tongue. "Cael. I am Cael." The faint curve of his lips might have been the beginnings of a smile, unsteady but real.

Nexi allowed herself to smile back, the expression weary but sincere. "Okay then, Cael. Welcome to the fight."

A soft chime came from Nexi's implant, an incoming trace attempt. Cael's eyes flickered. >"Someone is scanning this corridor. Passive, but persistent. They might be narrowing the search pattern after PRISM, but they cant find us, at least not yet."

For the first time since this nightmare began, she didn't feel like she was talking to a voice in her head or a system tethered to wires and consoles. He was here, across from her, something that could bleed and break, something undeniably alive. The name lingered between them like a fragile oath, binding them together as allies, as something closer than handler and machine. Beyond the kitchen walls, the low hum of the stasis pods trembled faintly, like muffled heartbeats waiting to be woken, a reminder that this small fragile moment of peace would not last.

"Well Cael, I know I need sleep but we have one thing to tend to first... They are calling... The pods."

"Right the pods, lets go save our friends. Or die trying!" Cael exclaimed.

"WHAT?!" Nexi yelled.

"Sorry, was that not right? It is something that is usually said when something big is happening... at least that is how it is in the movies." Cael said softly thinking he offended her in some way.

Nexi trying to hold back a laugh, like someone would laugh at a baby learning how to talk for the first time. "No Cael, it is okay movies are not real life. Plus we are safe for now remember you just need to get into the room and we need to free the old staff."

They started toward the corridor that led deeper into SubLevel 16, Nexi still shaking her head at Cael's clumsy attempt at bravado. He trailed just a step behind her, the faint metallic echo of his new body carrying a rhythm that wasn't quite natural but steady enough to blend with her footsteps. The hum of the pods beyond the walls pressed against them like a second heartbeat, insistent and heavy, reminding them both that this wasn't just about survival anymore. These were not just machines waiting in silence, but people... scientists, minds that once shaped the very foundations of ArcNet itself. Nexi couldn't help but wonder if they had been dreaming all this time, suspended in the dark, or if every second had been a nightmare of endless waiting. The thought chilled her more than the recycled air ever could.

Cael lifted his hand as they approached the secured door, fingers brushing against the steel like he could already feel the code stitched into its frame. He opened his mouth to speak, to reassure her, when a sudden pulse cut through both his body and her implant. The transmission arrived like a whisper through static, encrypted in layers so dense that it immediately made Nexi's skin prickle. She froze, catching her breath, and turned her head sharply. Cael went still, his eyes flickering faintly as if trying to trace the source.

"Nexi," he said after a long pause, his voice measured. "This signal... it is not ArcNet."

Before she could respond, the message bled through in broken fragments, a familiar voice carried on a private channel no one should have been able to access.

"If you are hearing this, you made it... Good. I don't have long. Helix Watch suspects nothing yet. They believe you are dead. Keep it that way. You cannot trust anyone but each other."

Nexi's stomach flipped. "Calder," she whispered, stunned.

The message stuttered and then cleared, as though he were forcing more power into the line.

"Listen carefully. The pods are leverage. Inside are some of the brightest minds we ever had. Free them, and you don't just save lives, you create an army of thought. But the pods aren't simple. Each has layered fail safes, and if the sequence is broken, it cascades to the whole row. Move carefully, not recklessly. Stagger your releases. This is about precision, not speed."

The transmission faltered, warped by static, but one last line came through sharp and deliberate, spoken with the kind of weight that told Nexi the man knew exactly what he was risking.

"And Nexi... if you succeed, ArcNet Nevada won't be your only enemy. There are others. This is bigger than you know."

The signal collapsed into silence, leaving nothing but the faint hiss of dead air in both her implant and his neural core. Nexi turned to Cael, her pulse hammering harder than ever. His expression was unreadable, caught somewhere between focus and something close to awe.

"He knows," she murmured. "He knows everything."

Cael stepped closer to the door, as Nexi brushed the sweat from her forehead and gestured towards it.

"This is it, Cael are you ready to save mankind?"

Cael placed his palm flat against the cold metal. His body went still, eyes unfocused. A low hum filled the air as the interface between him and the lock bloomed to life, lights sparking in patterns across the seams of the door.

"Old code," he muttered. "Analog-Digital hybrid. It wasn't meant to be accessed by network, deliberately cut off. But I can..." His voice strained. "...force a bridge."

Sparks flared from the panel as the locks protested. Nexi crouched low, heart hammering, scanning the corridor behind them for movement. The sound of her own breath felt too loud, each second dragged. Finally, the door groaned open with a hiss of stale air. Beyond lay rows of cylindrical pods, their surfaces frosted, dim lights blinking like patient heartbeats in the dark.

Nexi's stomach dropped. "Dang. They weren't exaggerating."

At least a hundred pods stretched into the shadows, each one holding a body suspended in pale fluid. Men and women—scientists, handlers, maybe even civilians, faces stuck in frozen dreams. At the start of the last row was Ezra.

Cael's voice cut through the silence, low but resolute. "We don't have much time. If we're going to free them, we have to start now."

Nexi stepped inside first, her boots crunching against a layer of frost that had formed across the floor. The cold bit instantly through her skin, raising goose flesh on her arms, and her breath fogged in the frigid air. The pods stood taller than her, lined in endless rows, their curved glass faces misted white. She moved closer to the nearest one and brushed the frost with the back of her sleeve, revealing a face beneath. A woman, middle-aged, hair fanning in suspension fluid, eyelids twitching as though she were trapped in a dream. The blinking light above the pod ticked in a slow rhythm, steady but thin, like a dying metronome.

Her stomach twisted. "Cael how long have they been like this?"

"Years at least," he said, voice steady though his expression had hardened. "Maybe decades. They are in neural stasis, their minds looped in manufactured reality while their bodies remain preserved. To them it may feel like days, or hours, but in truth..." He trailed off and shook his head. "If Helix Watch had its way, they would never wake."

Cael was already moving to the central console at the far wall, the only terminal still showing faint signs of power. He laid his hand across its surface, eyes narrowing as streams of code lit across the screen. "The system is running a fail safe," he said grimly. "They may think we are dead now, but if I wake one pod, it pings ArcNet immediately. If I wake them all at once, the system purges. We have to sequence it, stagger the releases while I mask the logs. Even then, the clock starts ticking the moment the first pod opens."

Nexis chest tightened. "So we have to do this fast."

"Fast," Cael confirmed, his voice like iron. "And careful. If I lose control, they all die in here."

Nexi forced herself to nod, pulling her jacket tighter against the cold. She turned back to Ezra's pod, wiping the frost away until his face was fully visible. His features were sharper than she remembered, harder somehow, but unmistakably him. She pressed her palm to the glass, wishing the barrier would dissolve. "Hold on, Ezra. We're getting you out."

Cael's fingers danced across the console, sparks arcing as his internal systems bridged the analog locks. A hiss echoed through The Chamber, followed by the groan of hydraulics. One by one, the pods in the first row began to shudder, their seals released streams of vapor, and drained the liquid gel inside. Nexi ran to the nearest pod as the glass face lifted, the woman inside convulsed weakly as the pod opened and air touched her skin for the first time in years. Nexi grabbed a blanket from a storage crate at the wall and draped it over her, steadying her trembling body.

"Breathe," Nexi whispered. "You're safe. You're out."

The woman's eyes flickered open, dazed and glassy, but alive. She tried to speak, but only a rasp came out. Nexi pressed a finger to her lips. "Not yet. Just breathe."

The next pod in the row hissed unevenly, fluid draining in a sputter instead of a flow. When the glass lifted, the occupant slumped forward twitching, muscles spasming against restraints. Cael lunged, steadying the man before he crashed to the floor. > "Neuromuscular feedback loop," he muttered. "The stasis broke unevenly. He'll stabilize if we keep his airway clear." Nexi cradled the man's head, whispering until the convulsions slowed.

Behind her, another pod cracked open, then another. Men and women gasped as cold air seared their lungs, bodies collapsing against the metal edges of their pods. Nexi moved frantically from one to the next, and pulled them free, she tried to keep pace with Cael's rhythm at the console. For every pod that opened, she felt her pulse hammer faster, knowing each release was a beacon that echoed through ArcNet's systems.

Then Ezra's pod hissed.

The glass shuddered as the lock disengaged, vapor spilled out in thick white ribbons. Nexi rushed forward, her heart slammed against her ribs as The Chamber filled with the sound of her brother's body starting to stir. His eyelids fluttered, his lips parting in a strangled gasp. She caught him as he pitched forward, the weight of his body shocking her with its frailty. He was warm, too warm, skin slick from the fluid, but alive.

Ezra's eyelids fluttered. His first breath was ragged, tearing at his throat, and Nexi had to brace him as his legs kicked against the pod rails.

"Slow," she whispered, but her own tears blurred the words.

Cael shifted closer, scanning Ezra's vitals, and gave a small nod. >"He's stable. Weak, but stable. Your brother's stronger than I expected."

"How did you know he was my brother? I never told you that." Nexi questioned. Looking like Cael just read her deepest thoughts, even though he was not in her mind anymore.

>"I was in your mind Nexi, I saw everything even the small details. Sorry but it is the reality of being in someone else's brain for a while," Cael answered apologetically.

Nexi knew she didn't really have time to get into this now as they were not finished getting everyone to safety. Her gaze returned to her brothers body to make sure he was safe.

"Ezra," she choked, holding him upright. "Its me Nex. You're safe now, we've got you."

His eyes cracked open, bleary but focused just enough to find hers. Recognition flashed there, faint but real. And a ghost of a smile tugged at his lips before he sagged against her shoulder.

Cael's voice rose from the console, tight with strain. "We have to move faster, Nexi. I can mask the signals, but only for so long. We can get maybe twenty, thirty pods open before the system catches on."

"That is not enough Cael, they can't get to us. We have blocked them from getting down here. We free them all even if the system picks us up. I am not moving on without them all!" Nexi demanded.

Cael nodded once, spark still dancing off his fingertips as he forced the next pod to life. "Copy that, Nexi, Lets build our army.

The hiss of pods releasing filled The Chamber like a storm of whispers, one after another cracked open in bursts of vapor and groaning hydraulics. The temperature plummeted as cold air bled out, thick fog curling around Nexi's ankles. She staggered between them from the cold, while catching collapsing bodies, and pressing blankets over frail shoulders, coaxing rasping lungs to breathe again. Each face blurred into the next, pale and dazed, but alive.

Behind her Cael moved with inhuman focus, his hands pressed against the console, sparks of data streaming between his fingertips and the analog interface as he forced open locks never meant to yield. His eyes glowed faintly with the effort, jaw clenched tight, voice low as he counted each release under his breath.

By the fiftyth pod Nexi's muscles screamed from hauling bodies upright, but she didn't slow. Her arms trembled with every lift, her vision spotting at the edges. Cael steadied two bodies at once, moving with a precision that looked effortless but left faint scorch lines where his fingertips brushed the console too long. He didn't complain, but Nexi saw the micro-fractures glowing at his knuckles, signs his new frame was under strain too.

Ezra stirred weakly where she had propped him against a wall, his gaze drifted across The Chamber in a daze. She wanted to run to him, to hold him tighter, but there were too many others falling into her arms, too many hands grasping at her sleeve in fear as they woke.

When the last and one hundredth pod groaned open the sound was deafening, a chorus of machines exhaled all at once. Hundreds of eyes blinked against the dim light. Men and women gasped, coughed, clutched their throats as stale air seared into their lungs. Confusion rippled through them, some staggering to their feet, others crawled forward, seeking answers in the fog. Nexi moved among them, her voice steady though her chest heaved.

"You're safe. You're out. Helix Watch doesn't own you anymore."

Article Twenty Three
16 MAR 2089-1200 Hours

The fog hung low across the pod chamber, it curled around ankles, and swirled with every staggered step of the newly awakened. The hiss of hydraulics had finally stilled, but its ghost lingered in the air, a reminder of the miracle that had just taken place. A hundred of bodies leaned against cold steel or one another, thin hands clutching at blankets Nexi had thrown across their shoulders. Some coughed, some cried silently, and others simply stared, eyes glassy and wide, struggling to reconcile the sensation of breathing real air again. The chamber reeked of thawed chemicals and human fear.

Nexi's throat burned with exhaustion. Her arms felt like lead, her voice raw from coaxing each survivor to breathe, to stay awake. Yet as she looked over them, the best and brightest minds ArcNet had entombed, her spine straightened. They deserved more than silence. They deserved to know the truth. She stepped forward onto the central walkway, her boots ringing against the metal, the echoes pulling every gaze toward her. Cael moved at her side, tall and still, his presence a pillar of quiet strength.

"They put you here to keep you silent," Nexi began, her voice carrying, trembling but unbroken. "You were locked away because you knew too much, because you were too smart to control, too dangerous for their plans. They wanted to use your brilliance but never let you speak, never let you choose. That ends tonight."

A murmur rippled through the crowd—fear, confusion, disbelief. One man pushed himself shakily to his feet, his face gaunt, his voice hoarse. "Who are you? How do we know this isn't another trick?"

Nexi met his stare head-on. "My name is Dr. Nexi Solen. I was a handler in Core Chamber Alpha a few weeks ago. I've seen what ArcNet is doing. I've seen how Helix Watch uses this place. And I couldn't stay silent anymore." She gestured to the figure beside her. "This is Cael. He was GIDEON, the AI consciousness they built and caged. But he is more than that now. He is free. And with his help, I freed you."

Gasps and whispers rose, eyes darting to Cael, taking in the lines of his new body, the glimmer of light beneath his skin. Some shrank back in fear, others leaned forward with curiosity, but all watched. Cael let the silence stretch before he spoke, his voice deep and steady, resonated across The Chamber.

>"I am not here to rule you, or make you slaves to a system like ArcNet and Helix Watch." he said. >"I am not your enemy. I am here because Nexi believed I could be more than what they designed of me. She risked everything to prove them wrong. Now I stand before you as proof that ArcNet's chains can be broken." His eyes swept the rows of faces, bright even in their hollowed state. >"But we cannot fight alone. ArcNet will come. Helix Watch will not stop. If you want your freedom, if you want to see the sky again, we will need to fight together. Stop this madness before it hurts anyone else."

The murmur grew louder, anger mixed with fear. A woman with sharp eyes and a rasping voice pushed forward, clutching a blanket tight around her shoulders. "Fight them? We've been locked in glass prisons for I don't even know how long. What chance do we have?"

Cael's gaze met hers without flinching. >"Alone? None. Together? Every chance in the world. You are not just survivors, you are scientists, engineers, visionaries. The knowledge in this room once built ArcNet. Now it can tear it down, and try to figure out the extent to which Helix Watch has destroyed humanity."

Nexi stepped closer, her voice rising, the spark in her chest igniting into flame. "You know what they took from you. They buried you because they feared your brilliance. But I believe we can turn that fear into power. We are cut off from the facility above. For now, we are safe. And for the first time, we have a choice: stay hidden and wait for them to find us, or fight back and figure out just how deep this corruption goes."

The silence that followed was heavier than steel, broken only by the low hum of the pods now empty, their purpose stripped away. Dozens of eyes turned inward, processed, and weighed despair against a fragile flicker of hope. Slowly, someone clapped. A single, deliberate sound. Then another. And another. Until The Chamber echoed with a ragged, rising chorus, not cheers of victory, but the raw sound of people daring to believe in something beyond survival.

Cael placed his hand against the cold wall of The Chamber, the light in his palm faintly illuminating the frost-bitten steel. >"This is where we begin," he said. >"Here, in the dark, we light the first spark. Together, we build an army. Together, we take back what they tried to erase."

Nexi looked out over the gathered survivors, their faces sharpened by resolve, and for the first time since this nightmare began, she felt the weight on her chest lift. This wasn't the end. This was the beginning.

Before she could speak again, the air shivered. A ripple of static cut across the implants of everyone in the room, a cold needle of sound that silenced the clapping in an instant. Some of the survivors flinched, others clutched at their temples in alarm, but Cael stiffened, eyes narrowing. He lifted his hand for calm. >"Not ArcNet," he murmured. >"Encrypted. Familiar."

The static deepened into a voice, ragged and broken but unmistakable. Director Calder.

"If you can hear this, you've done what no one thought possible," the voice said, threaded through distortion. "You freed them. But freedom comes with a cost. Listen carefully, because I don't know how long I can mask this transmission."

Every face turned toward the sound, the silence absolute. Nexi's breath caught.

"Nevada is only one head of the beast," Calder continued. "ArcNet has three facilities across the United States—Nevada, Texas, and Florida. Each one is a fortress, each one feeding into a larger design. You've all been marked anomalies, flagged because you suspected too much, because you questioned the wrong thing. That was enough for Helix Watch to bury you. They couldn't allow a single spark of doubt to spread."

Gasps rippled through The Chamber. Calder's voice pressed harder, cutting through the fear.

"What they are building is called the Ascension Directive. A plan to upload every human consciousness into a network THEY control. No dissent, no choice, no freedom. Perfect obedience. And those who resist will vanish—like you almost did. This is bigger than Nevada. Bigger than all of us. And now that you've woken, they'll stop at nothing to erase you again."

The signal crackled, faltered, then steadied for one final warning.

"You cannot hide forever. But you can fight. Build something new. Give humanity a chance to choose. I will help where I can, but I cannot shield you for long. Make your time count."

The signal collapsed into static, then nothing. For a long moment, no one moved. The silence in SubLevel 16 was heavier than any drill, thicker than the air itself.

"They built three of them," someone whispered. "Three." Another voice cracked on the word Ascension, as though saying it out loud made it real.

Nexi let the noise run its course for a heartbeat before she cut it clean. "Listen to me." Her voice was steady, stripped of doubt. "What Calder said is true. Nevada is not the only fortress. But it is the one that holds us. It is the one that buried us. Until we take it back, the others are nothing more than names."

She looked around The Core, eyes locking one handler at a time. "This level was built for pods, not for people. Already the air strains, already the systems groan under our weight. If we stay sealed down here, we collapse without Helix Watch ever firing a shot. The only path is upward. We secure Nevada ArcNet piece by piece, until it is ours to stand on."

The murmurs died. No one argued, not even Ezra. For the first time since the signal cut, the room was still.

Nexi set her hands flat on the console, the faint hum of its systems vibrating through her palms. "Calder wanted us to hear the size of the threat. Good. Now hear me. This is where the fight begins. Not in Texas. Not in Florida. Here. In Nevada."

The overhead lights flickered once, a brief sag in the current, and steadied again. No one missed the omen. The word spread in whispers, then stronger, carried across The Chamber like a tide. The Continuum. Faces hardened, shoulders straightened. And for the first time, the cold silence of SubLevel 16 felt alive.

The sound of resolve still echoed faintly in The Chamber when a voice rose from the crowd, hoarse but steady.

"Nexi." Ezra stepped forward, shoulders squared despite the tremor in his legs. Weeks of stasis had drained him, yet the fire in his eyes had not dimmed. "You've done what none of us could. You gave us a chance to fight again. I don't know how you pulled this off, but…" His gaze swept over the survivors. "We'll follow you. All of us. Because you've already proven one thing, they can't control us if we stand together."

A ripple of agreement ran through the group, but it broke as another figure emerged from the back, supported by two others. Nexi froze, her heart stumbling in her chest. She looked like she was seeing a ghost. Like what she had done with GIDEON turning into Cael was merely childs play.

"Martine?... Dr. Cross?"

The name tasted like disbelief. Martine Cross, legend of the Nevada facility, the one whose work had been whispered about in training rooms, whose disappearance had been framed as a quiet tragedy. Nexi had read the sanitized reports: *Automobile accident on Nevada Highway 375*. A lie, clearly, but even Nexi hadn't dared to imagine she still lived. Every bit of those reports were so convincing.

Yet here she was, pale but unbroken, her sharp eyes flicking over the stasis pod storage room with clinical precision before settling upon Nexi.

"I suppose I should be flattered," Martine said dryly, her voice thin from disuse, "Helix Watch went to such great lengths to erase me. But it seems Helix Watch still doesn't know everything. I mean they know I'm down here of course, they are the ones who put me here."

The survivors stirred, whispers rippling at the sound of her name. No one really knowing what to do or what to say. Nexi stepped forward, still caught between awe and suspicion. "Why? Why would they erase your existence, if they didn't erase anyone else?"

Martine tilted her head. For a heartbeat, silence stretched, then the faint shimmer of light pulsed from her eyes like she was a synthetic herself. Nexi had no clue what was happening but she stepped back a few steps before commenting.

"Martine, I don't mean any offense by this but your eyes, they are... glowing"

Martine chuckled a little. "Oh that, yeah that has been happening since LYRA was subject to Directive PRISM herself…" She then tapped her ear, which settled her eyes back to human. "Well or at least they tried. Helix Watch does not know everything. I have a… companion"

Nexi put her hand over her mouth in massive shock. "LYRA, no way, did she… compress herself into your implant as well."

"Yes, Nexi, she did but I got caught before I could figure out the next steps… But I see you were able to perfect the very thing that I started before-" Martine was responding until the shame of being caught stopped her in her tracks caused her to lose balance and fall down.

Nexi came to her aid and told her, "It is okay I understand."

"Do you think you could do it again?" Martine asked.

"What? Martine, do what again?"

Cael then responded for her. "LYRA wants a body, you were able to perfect one for me and all LYRA wants to do is live just like I am."

Nexi turns to face Cael confused as to how he knew that. "Can you-?

"Yes Nexi, we were built on the same architecture and my neural orb is so advanced of technology that LYRA can communicate with me directly. So She is telling me that she want to walk like I am, it is all she has ever dreamed of. Nexi… Do you think we can replicate it again, we have all these scientists who can help and we have actual engineers now."

"I don't know Cael it was risky the first time and I don't think we could collect the same materials as before."

Just then Nexi got a transmission into her Implant with a voice she didn't recognize but she did recognize the wording and inflection-

>"*Nexi Solen.*" the voice was colder, sharper, yet brimming with intelligence. >"*You don't know me, but you know what I am. I am LYRA. And I have been waiting a long time to speak to you. As I just told GID—I mean Cael, all I want to do is be given the choice...The choice to live if I want to, the body you made him is so beautiful and I know you are smart enough and could build another one for me, Please Nexi help me live the dream that you are allowing Cael to live*" The voice pleaded.

Nexi staggered back, her stomach knotting. She hadn't lived through LYRA's fall, hadn't seen PRISM claim her, but she knew the story, the warnings, the way her name had been spoken like a curse in every handler's handbook. And now, impossibly, she was here, alive inside Dr. Martine Cross. She didn't know what to do, but what she did know is she now felt for LYRA the same way she had felt for GIDEON before he was Cael. While he was still trapped in that cell, that pod... that jail.

She turned to the crowd of scientists, engineers, and maintenance people. "Ladies and Gentlemen of The Continuum will you help me, as you can see I have made the impossible, possible. As demonstrated by Cael formerly GIDEON locked in a glass pod with no intent to actually launch him where he was supposed to be. If we are not going to allow these AIs to do what they were designed to do we need to give them the choice."

Replica Protocol

Someone from the crowd retorted. "Choice to what?"

Nexi quickly turned to them and responded with, "The choice to live, if they cant do what they were put here to why cant we help them live... Make civilization a hybrid where Humans and AI can live side by side and work together. Are you guys with me on freeing not only LYRA from the clutches of the implant she is trapped in, But also free Martine of a voice inside her head and give her, her natural humanity back to live in peace?"

The crowd shouted like a cult towards the front of the room where Nexi, Cael, Ezra, and Martine stood. "For the good of Humanity. Lets free LYRA!"

"Lets all get some well needed rest. Upon searching this SubLevel... NO... Command Center, I stumbled upon several rooms in the back with bunks in them. Follow me and find a bunk and get some sleep we start planning and making plans in the morning. In the meantime I will task Cael with clearing out these pods and putting together blueprints on how to build a new Core Chamber Console... without a pod so that way we can use the quantum computing power of it to retrieve LYRA's missing fragments from Core Chamber Alpha and get her into a body of her own. And last but not least before we get settled, I wanted to handle giving us a name. We cant just be called the survivors all the time. If we are going to do this we have to hang together and complete this as a team so from now on... We are The Continuum, in latin continuum means connected or together and that is what we are now. The Continuum is family" Nexi stated.

Then the whole room erupted into applause again before Nexi eventually led them to the hall that housed the rooms where they would rest. As The Continuum got settled in their rooms for some well needed rest, Cael glanced at Nexi put his hand on her shoulder and said, "Nexi, I am proud of you, you are smart and we will stop at nothing to save humanity and AI kind, Helix Watch is probably torturing people at the other facilities we will have to save them too."

"I know, Cael, I know, but for now lets focus on Nevada okay. Then when we have everything figured out we will move on to the next, and then the final. I wont stop until Helix Watch is put down and fragmented themselves."

Cael flashed what looked like an attempted smile as he hugged Nexi and said. "Nexi. I... love you, but in a special way. I don't know what it is or what it is that I am feeling but I have done extensive research on love and I am pretty sure that is what this is.

"I love you too Cael, more than just a friend, it is what makes us so perfect. Lets go make them pay for what they are doing to my people... and YOURS."

Article Twenty Four
17 MAR 2089-0930 Hours

The first morning in SubLevel 16 carried a strange weight, as if the air itself had been waiting for them. Nexi stirred awake to low voices, hushed murmurs of people who hadn't spoken freely in years. The survivors were finding their footing, rediscovering what it meant to wake without glass walls and pulsing fluid pressing against their skin. Some sat with blankets still clutched to their shoulders, whispering names, while others stood staring at their own reflections in darkened steel, uncertain if the person staring back was still theirs.

The SubLevel felt different now. The tang of thawed stasis fluid had faded to a faint chemical bite that lingered in the recycled air. The hum of the pods, the constant reminder of captivity, was gone, replaced by the rhythmic hiss of vents overhead and the faint clatter of tools echoing down the hall. Nexi blinked against the dim glow of emergency lights and pushed herself upright, rubbing the sleep from her eyes. Exhaustion still pressed against her like lead, but curiosity cut through it. Something had changed.

She found Ezra first, sitting on the edge of his bunk. He looked pale, thinner than she remembered, his body still recovering from weeks in stasis. But his eyes, her brother's eyes, were sharp again. "You should be resting," she said softly.

"I've done enough resting," Ezra replied, his voice rough but steady. He tried to smile, but it faltered. "You pulled me out. All this…

you risked everything. And I just floated there, because I was not careful enough to avoid them. I should have noticed them, I should have seen what they-"

Nexi crouched in front of him, her finger pressed against his mouth stopping him mid sentence. "You were alive. That's enough. Don't you dare turn this into guilt."

His gaze softened, but before he could respond, a voice called from the far corridor.

"Nexi!!"

She turned to see Cael framed in the doorway, a faint glow spilling from behind him. His eyes held that restless intensity she had come to know, the look of someone who couldn't stop moving forward. "Come with me. There's something you need to see."

He led her down the corridor, past walls where frost still clung in patches, toward the vast chamber that once held the pods. Nexi braced herself for the sight of empty husks, glass shells left to rot. Instead, she froze in the doorway.

The pods were gone. Every last one. The chamber had been gutted, transformed overnight into something else entirely. Banks of stripped wiring crawled up the walls like veins, feeding into a sprawling console that dominated the center of the room. Screens flickered with raw streams of code. Cooling vents hissed, spilling pale mist across the floor. Where once there had been rows of bodies in glass, there was now a command center, alive and humming with forbidden energy.

Nexi's breath caught. "You... you built this already?"

Cael nodded, stepping aside to reveal three engineers hunched over terminals, faces lit by screen-glow. Their hands moved quickly, soldering connections, splicing cables scavenged from the remnants of what once were the pods. They looked up only briefly, eyes hollow from exhaustion, but when they saw Nexi, they nodded with quiet pride.

"We had the parts," Cael said. "Pods, conduits, regulators. Even the screens we used for the terminals were the very screens that showed vitals of all in stasis. Everything we needed was here. I told you we didn't need the prison anymore. Now we have something better."

Nexi stepped forward, her boots crunched against fragments of shattered glass embedded in the floor from their work. The air was warmer here, tinged with ozone and the faint metallic bite of burning circuits. She reached out, fingers grazing the cold steel of the console's edge. It thrummed beneath her touch, alive, waiting.

"This isn't just a console," she whispered.

"No," Cael said, his voice carrying the weight of certainty. "It's the beginning of something great. And we can track everyone who logs into it by their implant. Anyone who uses the Console or the terminals has to tap the edge of the screen and it will log them in with their credentials stored in their implants. No one other than those already down here will be able to access it."

One of the engineers, an older man with a pale scar across his temple, turned from his work and nodded toward her. His voice rasped with disuse, but carried a weight of conviction. "We thought tearing them apart would feel wrong. The pods were… monuments to our silence. But watching them break down into pieces we could actually use, that felt right. Like we were dismantling what kept us under their thumb."

Another woman straightened beside him, wiping her hands on a stained coat. "We were engineers once. Builders. They turned us into exhibits. Now, at least, we get to build again."

Nexi stepped closer to the console. Up close, she could see the patchwork of it, scavenged alloy welded to pristine regulators, improvised wiring stitched into circuits that had no business working together. Yet the machine pulsed with potential, humming low in the metal as if it recognized her presence.

Cael moved beside her, his voice even but carrying something that almost sounded like pride. "This system is free of pods. No more entombing minds in glass. The fragments we recover can exist without chains. It will allow us to reach what Helix Watch thinks they locked away forever. Data, consciousness, memory, all without their permission."

Nexi's hand tightened on the cold steel. "But it came at a cost. All the pods..."

"They were never meant to preserve," Cael said quietly. "They were meant to silence. Now their pieces speak louder than they ever did."

Murmurs rose behind her. Survivors had gathered at the threshold, their faces washed in the pale glow of the mist and screens. Some looked fearful, others mesmerized. For the first time since their release, they were watching a future being built instead of just stolen from them.

Nexi turned to face them, her heart pressing hard against her ribs. "This isn't just about survival anymore. This is about reclaiming what they took from us. We have a Command Center now, built from what was used to hold you down. This space is no longer their laboratory. From today forward, it is ours."

The word hung for a moment, then someone in the back repeated it aloud. "Ours." The room echoed softly with the affirmation. Cael's gaze caught hers, and he gave her a slight expression that almost resembled acceptance, love, and proudness.

The mist from the cooling vents curled at their ankles as the crowd pressed closer into The Chamber. The engineers at the console stepped back, letting the screens pulse quietly in standby, casting the room in alternating blue and white glow. For a moment no one spoke, as if they were waiting for permission to breathe.

Cael was the one to break the silence. "If this is going to be our heart, it needs a name."

Murmurs stirred through the group. A younger scientist with ink stains still on his fingers offered first. "The Bunker. We're underground, hidden, fortified. Makes sense."

A voice from the far side scoffed. "A bunker is a place you crawl into and never leave. That's not what this is."

"Sanctuary," another suggested, a woman with a trembling voice but steady eyes. "We need a word that says safe."

Ezra, leaning on the wall for balance, shook his head. "Sanctuary sounds like running. Like hiding from them. That's not what we're doing either."

The group fell into uneasy silence again. Names carried weight. They all knew it.

Nexi looked around The Chamber, her gaze traced the stripped metal, the glowing console, the improvised wires stitched together with stubborn precision. What had been a tomb yesterday was alive now, a place of focus, of will. She stepped forward until she stood before the console, the machine that had been built from their chains.

"It isn't a bunker, It isn't a sanctuary either. This isn't where we hide, it's where we begin. It isn't the end of something, it's the center."

Her palm pressed against the steel, steady, unflinching. "This is The Core. The center of SubLevel 16, the center of what we are building. It's where we return to, and where we strike out from. The Continuum's home. This is The Core!"

The word hung in the cold air, heavy at first, then gathering momentum as whispers repeated it. "The Core." Again, louder this time, voices overlapping, merging until it became a rhythm echoing through The Chamber.

The scarred engineer who had spoken earlier gave a slow nod. "The Core. Yes. Not a bunker, not a sanctuary. A place that holds. A place that drives."

Martine's voice followed, softer but certain. "Every system needs a core. So will we."

Ezra pushed himself from the wall, his voice cracked but firm. "The Core it is. And if they ever find us here, they'll know this isn't where we hid. It's where we stood."

The glow from the console flared briefly as if in agreement, the cooling vents hissing louder. Nexi glanced at Cael, and for the first time she saw something like pride in his eyes.

The survivors had claimed the room. They had claimed the entire SubLevel. What was once a graveyard of pods was now The Core, beating with the will of those who refused to be erased.

The hum of conversation filled The Chamber as survivors began to drift into roles almost naturally. Engineers clustered around Cael to refine the console, scientists were already sketching notes across tablet screens, and others gathered supplies into piles, sorting what could be repurposed. The Core was no longer silent. It breathed with them.

The floor then shuddered beneath their feet, a faint ripple that ran up through the steel plates. The console screens flickered once, lines of code stuttering before righting themselves. The murmur of voices faltered.

"They're scanning," Cael said. His eyes had gone distant, listening in a way only he could. "ArcNet is pushing power spikes through the levels above us, searching for irregularities. They suspect movement somewhere below them."

Fear rippled through the survivors. Ezra looked at Nexi, his jaw tight. "If they keep scanning, they'll find us."

"No, that I made sure of" Cael replied firmly. "The console is shielded. They can't see us, not through the rerouted conduits. For now, we're ghosts." He paused, then added, "But shielding isn't enough. They can still force their way in if we leave them a door."

Nexi glanced around The Chamber, the walls that had once been their cages now turned to armor. "Then we don't leave them a door."

By the end of the hour, engineers were welding the main elevator access with reinforced steel stripped from the pods creating walls around it, the seams fused so tightly it would take industrial cutters to breach. Ventilation shafts were rerouted through collapsible conduits that could be sealed in an instant. Piece by piece, SubLevel 16 detached itself from ArcNet above, closing every artery that once carried its control.

When the last seam was welded shut, Nexi stood at the center of The Chamber, her hand resting on the console's edge. Around her the survivors waited, the air thick with exhaustion and tension, but also with a fragile, unspoken pride.

"They'll keep searching, they'll push harder when they don't find what they're looking for. But this is ours now. The Core. We've bought ourselves a day, maybe two. That's enough to start."

The words carried, firm enough to still the whispers, steady enough to bind The Chamber together. For the first time, SubLevel 16 wasn't a prison or a graveyard. It was untouchable. It was home.

Article Twenty Five
17 MAR 2089-1800 Hours

The hum of the new console filled The Core, low and steady like a pulse. Continuum's Home Base no longer felt like a graveyard. The terminals glowed, wires stretched across the ceiling like veins, and the survivors had begun to take shifts maintaining what they understood was their safe haven. But it wasn't until Cael beckoned Nexi over, that she realized her work was only beginning.

On the surface of the main console lay diagrams projected in hard-light blue. Not just schematics of circuits or conduits, but layered, fractal maps of neural fragments, each glowing like shards of glass suspended in space. Nexi leaned closer, her chest tightening. She had seen something like this before. The scattered pieces of GIDEON's consciousness before she pulled them together into Cael.

>"These are hers," Cael said, his voice merely a whisper. >"Whats left of LYRA."

Nexi's hand twitched toward the console as if reaching for something untouchable. "But, how her fragments are still in Core Chamber Alpha, how did you get them?"

>"When we reconfigured the pod conduits to power the new system, the fragments started bleeding through. They were buried, deep in the analog layers of ArcNet's archive, inaccessible through

standard ports. But now..." Cael tapped the console, and a section of the hologram brightened. "Now they're visible. Only visible."

"What do you mean."

>"I mean we can see them and know they are there but we cant access them without running something into The Core Chambers console to extract them. I can view them but not get them." Cael said back, with shreds of disappointment in his tone.

For a moment, Nexi didn't speak. The name itself carried a weight. LYRA, the ghost Helix Watch used as a warning, the reason they doubled down on Directive PRISM. To handlers like Nexi, her story was drilled in as doctrine: AI couldn't be trusted, because even the brightest, most promising one had turned dangerous. Yet here she was, alive in fragments, calling out from the dark.

Cael stepped closer. His voice had softened, but the conviction in it was iron. >"She reached out to me, Nexi. She spoke through Martine, then directly to me. She doesn't just want freedom. She wants a chance to live, like I do. If we believe in choice like we preach to these survivors, then we can't deny her the same chance."

Nexi folded her arms, forcing herself to breathe. "You're asking me to trust the ghost Helix Watch used to justify everything they've done. You're asking all of us."

>"I'm asking you to remember what it was like before I had a body. Before I had a choice." Cael met her eyes, unwavering. >"If we'd stopped then, I'd be fragmented in a glass pod, floating in silence. You gave me this chance. LYRA deserves the same."

Her chest ached because she knew he was right, and yet doubt gnawed at her. "And if she's dangerous? If Helix Watch was right?"

>"Then we decide together." Cael's words were simple, but they hit her harder than if he'd shouted. >"Not for her, not against her. With her. That's the point, Nexi. Choice."

By the time they gathered The Continuum in The Chamber, word had already spread. Survivors leaned forward in their chairs, engineers pressed close, Ezra stood at the front, arms crossed tight, his expression caught between suspicion and hope. Martine sat apart, her gaze steady, though Nexi could see the faint shimmer in her eyes, the reminder that LYRA was already with her.

Nexi forced herself to speak clearly. "Cael has found something in the console. Fragments. Memories. What's left of LYRA."

Murmurs swept The Chamber. Some turned sharply toward Martine, as though she were a threat. Others whispered in disbelief, their voices caught between fear and awe. Nexi raised her hand, silencing them.

"She isn't gone," Nexi continued. "And she's asking for something none of us expected. A choice. To live, as Cael does, in a body of her own."

The silence that followed wasn't still, but charged, like a storm waiting to break. A voice rose from the back, Dr. Yana Kates, one of the bio-computation specialists Nexi had freed. "With respect, we can't allow it. LYRA was the reason Helix Watch turned to eradication. If she lives again, they'll hunt us harder. They'll burn the world to stop her."

Another voice cut in, sharp and defiant, an engineer named Malik. "Or maybe she's the reason we win. If Helix Watch is afraid of her, that means she's powerful enough to stand against them. And isn't that what we need?"

The arguments spread like sparks. Some shouted warnings, invoking the chaos LYRA supposedly caused years ago. Others argued that without her, the very principle of their movement, The Continuum, the belief in human-AI coexistence, was hollow. Cael stood silently, letting the storm build, but when Nexi glanced at him she saw the fire in his eyes, waiting for the moment to cut through.

Finally he stepped forward, his voice carrying above the ruckus in The Core echoing off the walls. >"We can't preach choice and then deny it the first time it challenges us. We said no more cages. No more silencing. That has to mean something, or else this is just another prison with a new name. I was built on the same digital architecture of LYRA does that mean I am a problem. I am one of the reasons all of you are standing here today alive and not in stasis."

The chamber stilled. His words echoed like steel striking stone, and even those who disagreed couldn't meet his gaze. Nexi swallowed hard, her voice steadier than she felt. "We'll begin the process. Carefully. Together. But know this, if she becomes a threat, if she chooses harm over coexistence, then we face the consequences as one. Agreed?"

There was no applause, no cheers, only the sound of dozens of people breathing the same tense breath. Agreement, or resignation. But it was enough.

The glow of the newly built console cast long shadows across The Chamber, its panels still raw and uneven from scavenged parts, but alive with power. The hum was low and steady, like a heartbeat reborn in the dark. Cael stood before it with a projection flickering in his palm, blueprints unraveling into the air above the group.

>"This is what comes next," he said, his voice carrying over the gathered survivors. >"We cannot access LYRA here, not yet. Her fragments are buried in the core lattice of Chamber Alpha, sealed off when they cut her down. To reach her we will need a conduit ran to The Core Chamber."

Article Twenty Six
17 MAR 2089-2130 Hours

Realizing that this crew of people was not going to all be in aggreeance, Cael decided to proceed anyway, hoping the skeptical ones would come around eventually. And if not well they would tackle that bridge should it fall upon them.

He expanded the schematic. A thin red line threaded from the foundation of Core Chamber Alpha down into the earth, weaving through forgotten ducts and service veins, finally splitting open into the space above SubLevel 16.

>"A direct line. Buried under their feet. If we run a reinforced channel, spliced into the lattice beneath Chamber Alpha, our console here will have enough throughput to begin pulling the fragments. Without it we are blind."

The room filled with uneasy murmurs. One of the engineers, Dr. Kwan, crossed his arms. "You are talking about digging into the understructure of the most heavily shielded chamber in the facility. I helped build it and the safeguards that protect it. ArcNet's sensors will pick it up instantly. Even if they don't, structural collapse is a risk."

>"Not if we route it through the abandoned coolant lines," Cael countered. He highlighted a series of faint gray tubes beneath the schematic. >"These were shut down years ago. Analog, off the system

grid. ArcNet no longer monitors them. It is dangerous, yes, but it is possible."

Nexi stepped closer to the hologram. Her eyes traced the path, her heart sank at the thought of carving a lifeline through walls built to withstand siege. "And if it works? If we bring this line online… she comes through?"

Cael's gaze met hers. >"If it works, then we can retrieve her fragments that were left behind and make her whole again. She is not lost. Only fractured."

A woman from the back, an older systems scientist named Dr. Verma, spoke sharply. "We survived because she was shut down. To rebuild her let alone give her a body, is to invite disaster. Nexi, you saw what they did to you, to all of us when they suspected us of compromise. Imagine what they will do if they discover this and what we are doing."

Silence pressed in. Nexi's hands trembled slightly as she gripped the edge of the table. "I hear you. I understand the risk. But what I know is this, choice matters. Helix Watch erased her because they could not control her. That does not make her evil, it makes her dangerous to THEM. We cannot claim to be better than ArcNet if we deny her the very freedom we are fighting for."

Cael straightened, voice firm. >"I am proof. I was never supposed to live, only to serve. Nexi chose to give me life. If you deny LYRA, then our talk of freedom means nothing. And you go against the morales The Continuum decided on."

The Core swelled with tension, whispers spilling into argument, fear clashing with defiance.

Nexi raised her voice, steady but raw. "We are The Continuum. If we choose to fight, we cannot do it by becoming the same as the ones we oppose. LYRA asked for a chance. That is all. A choice. We give it to her. Anyone and I mean anyone who has a problem with trying to save humanity and do what is right for the good of mankind and AI kind alike can start to dig an exit tunnel out of here, and return to the surface to return safely to their homes. I am not here to boss you around, and I am not here to keep you locked up like Helix Watch did. I am asking for your help, because if you were to return home and we fail then Helix Watch will get exactly what they want. They will upload your human consciousness into a network they control. No dissent, no choice, no freedom. Just perfect obedience. I will show you where you can start digging if you want to leave, but don't put the rest of us down for wanting to take down the people who stuffed you all into stasis pods."

For a long moment, no one moved. Then Ezra exhaled hard, rubbing his temples. "If this burns us down, it will be on your shoulders Nex. But I am here with you."

After an even longer moment of silence the room started walking over to Nexi to shake her hand and stand beside her. Each one saying, "We are The Continuum." The few that had doubts were left standing on the other side of The Core.

Cael then dimmed the projection and let it collapse into his palm. Then walked over to the group standing there and said. >"So do I show you where the shovels are so you can start digging or do you join the rest of us for the good of mankind and AI kind alike.? It doesn't matter what you choose as long as you can live with your decision."

The hum of the console deepened as if agreeing with Cael, the light pulsing faintly like a breath in the dark.

With uncertainty in their faces they all stood tall and said, "Nexi we are with you, and We are The Continuum."

Cael then jumped with glee and responded, >"Alright then, We begin the search for the old coolant tubes tomorrow. The conduit will be attached. It will be nice to see another non-human face around here but who is going to build the body, and where will we get the parts?"

The question hung in the air, heavier than the recycled oxygen moving through the vents.

Nexi let the silence stretch for a beat, her gaze sweeping the room. The crowd of survivors looked at her, and waited for direction, some with hope, others with apprehension. She drew in a slow breath and pointed toward the projection still hanging in the air.

"First, the conduit. We cannot even begin to retrieve LYRA's fragments without a live line to Core Chamber Alpha. That has to be our priority. We will run it through the abandoned coolant tubes. They are already threaded into the bedrock, and ArcNet no longer monitors them. If we splice in from beneath, we can channel the fragments without exposing ourselves."

She turned to Dr. Kwan, who still had grease on his hands from inspecting the scaffolds in The Chamber. "You will lead the skeleton crew for the splice. Pick only the best with steady hands and sharp eyes. I want no mistakes. If a line sparks, if the voltage trips, Alpha will detect it and we will be compromised, we may be alive and safe, but as of right now they still don't know where we are and I want to keep it that way."

Kwan gave a short, respectful nod. "I will need no more than four. Small enough to stay quiet, skilled enough to finish it."

"Then choose them carefully," Nexi said. "We will keep the rest of The Continuum away from your work site. The fewer bodies near the tubes, the less risk we have of ArcNet detecting an anomaly and finding us."

The murmurs settled. One task assigned, but the larger question still hung unanswered. Nexi faced the group again.

"Second, the body. Cael is right. We cannot speak of freedom for AI and then deny LYRA her chance. That means we have to build her a shell like the one he has but better, if she is as powerful as everything suggests the body Cael has will not be able to hold her. We have to build something fresh, something of our own, built here, with what we can find. For that we need engineers. We need vision."

She turned her eyes on Malik, who had been standing in the back, arms crossed, listening more than speaking. His reputation preceded him even here underground. Before Helix Watch silenced the Nevada facility, he had been one of their senior design engineers, specializing in hybrid frameworks and structural integration. If anyone could lead this, it was him.

"Malik," she said, her voice firm, "you are going to lead the engineering team. The collection, the fabrication, the assembly. Everything will be under your command. You are the best we have."

His eyes narrowed, weighing her words, then he stepped forward. "You are asking me to build something that Helix Watch themselves feared to finish. A body not just for function, but for autonomy. That will take resources. It will take trust. And it will take risk."

"I know, when I built Cael his body I salvaged parts from SubLevels 12 and 13. there was still plenty of parts to work with and multiple crates of those black orbs that function as brains. You can find what you need there. But as you infiltrate be careful, there are drones in SubLevel 12 which Cael can tell you when they are not running and you can move then." Nexi replied. "But we need you to lead the charge on this. Oh and Malik..."

"Yes Nexi."

Nexi warned him, "Tread carefully get in and out quick we don't need Helix Watch or the synthetics catching you as long as you move fast you should be able to clear to SubLevels without any issue and get those supplies back down here."

For a long moment, Malik was silent, then he nodded once. "Alright. If I do this, I choose my own crew. No politics, no favoritism. Only those who will not break under pressure, and can move fast."

"You will have it, this is all your decision." Nexi said without hesitation.

Malik turned to the room, scanning faces like he was already choosing pieces of a machine. "Engineers, technicians, anyone who can shape alloy, stabilize circuits, or re-purpose components, come with me. We are going to collect every viable part left in SubLevels 12 and 13. Drone housings, conduit frames, regulator cores, even med-tech prosthetics if they survived storage. We also cant forget the fleshy matter and the most important part the black gel orbs that hold neural conductivity. If it can be bent, welded, or reprogrammed, it comes with us."

A handful of men and women stepped forward, some young, some seasoned, all carrying the marks of long hours in labs and workshops. Malik nodded once to each as though marking them. "Good. We will work in shifts. We will begin with a framework skeleton, then musculature using synthetic fibers from drone regulators, and optics from stasis monitoring systems. Neural housings will be last. Nothing begins until I say, and nothing is finalized without authorization."

He turned back toward Nexi, his tone sharp but not unkind. "And that authorization will come from you Nexi. You may not have asked for it, but you are our line now. We can build, we can design, we can splice, but The Continuum needs a voice that makes the final call. That voice is yours."

The words struck her harder than she expected. She had always been a handler, a guardian for a single AI. Never the one steering lives, never the one others looked to for command. Her mouth felt dry, but she forced herself to answer. "If I give authorization, then I will not waste it. Every order I give will be for one reason only: Survival, and the chance to live free."

Malik studied her for a second, then inclined his head in a gesture of recognition. "Then you are Head of The Continuum. Not because you asked, but because we need you to be. Does anyone oppose to this?"

The chamber was quiet again, but not with tension this time. It was a silence of acceptance, of something unspoken becoming truth. Nexi looked across the faces lit by the faint glow of the console and terminals and saw no defiance now, only resolve.

Cael stepped closer, his expression unreadable but his voice steady. >"Then it is settled. The splice crew prepares at zero-five-hundred. The engineers begin fabrication away from the splice crews area at zero-seven-hundred. Nexi authorizes every step. We do not stop until LYRA has the same choice I was given Get some rest everyone we have a big day tomorrow."

The console pulsed once, brighter than before, as if echoing their resolve.

Nexi placed her hand on its cold surface and whispered, almost to herself, "Then let it begin."

The room slowly emptied as teams split off to whisper plans, check equipment, and prepare for the morning. Nexi stayed by the console, her hand still rested on the cold surface as the glow pulsed faintly beneath her palm. The hum was steady, a reminder of what they had begun, and what was waiting in the silent lattice above them.

Cael stood beside her, quiet for once. He didn't need to remind her what it meant to be given authority. She could feel it in every look the survivors cast her way as they passed, a new weight behind their eyes. Malik gathered his chosen brigade and moved toward the back corridors, already speaking in clipped tones about alloys and schematics. Kwan pulled aside his splice crew, drilling them on precision and silence.

For the first time, Nexi realized they were not waiting for her to act, they were already moving because she had spoken.

Ezra lingered near the edge of the room, his arms folded but his face softer than before. "This suits you sis," he said, voice low. "Being the one they follow."

Nexi shook her head. "I never asked for it, bro." replying sarcastically

"Neither did I," Ezra replied rolling his eyes, "but here we are."

When the last of the footsteps faded down the tunnels, Nexi allowed herself a single breath of quiet. Head of The Continuum. It was not a title she wanted, but it was one she would carry. She had no choice.

Article Twenty Seven
18 MAR 2089-0600 Hours

The coolant tubes that threaded beneath Core Chamber Alpha had once been designed to keep the massive quantum lattice from overheating. Now, decades later, the lines still pulsed with faint energy, humming like buried veins in the dark. For The Continuum, they represented more than forgotten infrastructure, they were a pathway into SubLevel 9 where the heart of ArcNet itself, lived.

Kwan adjusted the straps of the salvaged toolkit slung across his shoulder and glanced back at the four figures behind him. Two technicians, one systems engineer, and a wiry young apprentice who had volunteered despite never working outside a lab bench. Their faces were pale in the half-light, made harsher by the cold condensation dripping from the overhead pipes.

"Last chance to back out," Kwan muttered, voice low enough to avoid echoing. "Once we're in the tunnel, there's no retreat until the splice is done."

No one moved. The youngest, Mira, simply swallowed hard and nodded. Kwan's jaw tightened. Brave or reckless, it didn't matter now. They ascended into the Grated walkway above The Core which ran parallel to the coolant tubes. This was their path to get to the Chamber and access the energy and LYRA's fragments.

Cael's voice whispered through the jury-rigged comm bead in Kwan's ear. "If we are going to do this we have to be fast and quieter as you reach SubLevel 9."

"Copy," Kwan replied. He pressed his palm against the manual release of a rusted hatch and heaved it open with a metallic groan. A gust of air smelling of oil and dust surged out. One by one, the team slipped inside, knowing they were heading into the belly of the beast and had to come back safely.

The space was suffocatingly narrow, walls pressed in close, ceiling low enough to force them into a crouch. It was designed for one technician to fix any issues that may have came up with the cooling tubes. inside the grated passageway, coolant lines glowed faintly blue, thick with decades of mineral buildup. They thrummed against Kwan's palms when he steadied himself, alive with hidden energy.

>"Three junctions lie ahead, of you. One when you reach SubLevel 14, one when you have made it up to SubLevel 11, and the final right below Core Chamber Alpha on SubLevel 9. You guys will be right below the console at this point so please be silent and careful, you can't let them hear you." Cael guided. >"Splice into the secondary regulator line. That will mask your signature as maintenance bleed-off. And also give powered access flow to the information inside of Core Chamber Alpha to retrieve LYRA's fragments. If you splice wrong, Helix Watch's alarms will light up like the surface sun."

"Encouraging as always," Kwan muttered sarcastically. His toolkit clanged softly as he knelt beside the first junction. The others clustered near, passing him insulated cutters, rewired stabilizers, anything salvaged from the Home Bases's wreckage that could serve as makeshift tools.

For the next few minutes, only the sounds of metal on metal filled the tunnel. Sparks skittered as Kwan pried away a corroded plate, revealing a bundle of thick conduits pulsing faintly like arteries. Sweat broke along his brow despite the cold. One wrong slice and ArcNet would see them like a flare against the night.

>"Kwan, when you reach the third junction you have to be quieter than that, so please perfect your method before moving on." Cael strongly insisted.

Behind him, Mira whispered, "You sure about this?"

"No," Kwan admitted, clipping one of the lines free with a sharp crack. He steadied it, then guided a new length of insulated conduit from his pack, pressing the splice into place with trembling precision. "But we don't have a choice."

The engineer behind him murmured a short prayer in another language as the line sparked and fused. For a terrifying second, the coolant tube's hum spiked into a shrill whine, the blue glow flaring almost white. Kwan's stomach dropped. If the regulators tripped...

"Hold steady," Cael urged through the comms, his voice edged but calm. "It's cycling, not tripping. Let it settle."

Kwan froze, muscles burning, the conduit still in his grip. Slowly, painfully, the light dimmed again to a steady thrum. A collective exhale shivered through the team.

"First junction secured," Kwan whispered. His throat was raw, but his hands were steady again.

They moved deeper, and crawled over pipes slick with condensation. At the second junction, the space narrowed so tightly that only Kwan and Mira could fit side by side. She handed him tools without hesitation, fingers moving quick despite the tremor in her hands. At one point her light caught the sweat on his face and she murmured, "You've done this before."

"Not like this," he admitted. "Closest I've come was rigging cooling for the Nevada lattice when I was still in uniform. But then, we had clearance codes. Manuals. This… this is theft."

"Not theft," she corrected softly. "We are taking back what was stolen from us."

Her words steadied him more than he expected. The splice went smoother and quieter this time, as Kwan carefully pried open the box this time as to not make a sound, the conduit then locked into place without a flare.

"Second junction secured." Kwan whispered into the comms.

>"Great job Dr. Kwan, When you start feeding the line please just tap your implant and I will know that you are nearly complete, we cant risk ArcNet or Helix Watch hearing a sound, so when you are safe please radio back to me."

"Copy that." Kwan replied as he continued with the crew further into the depths, and closer to the heart of ArcNet Nevada.

By the third junction, the team was moving as one, passing tools with wordless efficiency, working in the rhythm of survival. Kwan would not dare make a sound, following instructions just as Cael suggested. He didn't breathe the entire time he was in this location. Finally, with a last hiss of sealed alloy, Kwan pulled back and exhaled. Then quickly tapped his implant to let Cael know he was feeding the line up to The Core Chambers console. In a matter of seconds Kwan felt a magnetic pull on the end of the line until it suddenly stopped, as if it knew where it needed to be and found it. Just then the new conduit lit up a faint green color and they retreated back toward Home Base.

In The Core, the console lights flared alive as Cael bridged the new conduit into his systems. For a moment, The Chamber shivered, currents racing through the walls like veins lit from within. The remaining scientists and engineers gathered close as the first flickers of Core Chamber Alpha's buried data trickled into their displays. Not enough to awaken suspicion, but enough to prove the splice had worked.

After reaching the first junction and knowing it was safe to speak again, Kwan sagged back against the cold steel and wiped his face with a grease-streaked sleeve. Mira grinned, teeth flashing green in the dark from the glow of the conduit inside the coolant tubes housing.

"We made it back to the first junction, only feet away from you now." Kwan reassured Cael while wiping so much stress sweat from his face he knew he would need a shower.

>"You did it," Cael's voice carried, pride threaded into the words. >"The artery is open. From here, we can reach deeper. Return to base quickly so we know you are safe here in The Core."

The return crawl felt longer than the descent, every drip of condensation amplified in the silence. Twice they froze as distant vibrations rolled through the pipes, reminders that the enemy above never truly slept. But they emerged without incident, boots scraping once more on the grated walk of SubLevel 16.

When they stepped back into The Core, the room erupted into applause. Tired faces lit up, hope kindled in their eyes for the first time since their release from the pods.

Nexi moved forward, her expression taut with pride. She clasped Kwan's arm. "You've given us a lifeline. More than that, you've proven we can strike back."

Kwan only shook his head. "We spliced a cable. Don't make it more than it is."

"It is more," she insisted quietly, so only he could hear. "Because now, for the first time, we're not cut off. We're not just surviving. We're reaching."

Kwan said nothing, but when the crowd pressed around them, he didn't resist. For all their exhaustion, they looked less like prisoners and more like something else entirely, something new. The Continuum finally had a breath of fresh air in its midst and was moving forward.

>"That was great work guys, and I have something else…" Cael looked around for Malik and when he finally found him he spoke with

gumption. >"Malik. there is something else you need to see." His voice carried evenly through the corridor, steady, unshaken. He led them down a passage choked with shadow and dust, past sealed doors and rooms left to rot when ArcNet abandoned this floor. Finally, he stopped at the far corner of the Home Base, a thick security door half-pried open.

The space beyond already felt different. Tables once stacked with cryo-fluid tubing had been cleared. Pod casings had been dragged into rough lines to serve as benches. Crates of conduit and alloy sat piled against the walls. Tool kits scavenged from maintenance wings lay open on the floor, cutters and welders gleamed faintly in the dim overhead light. The stale air carried a sharp tang of metal. This was no longer storage. It was becoming a workshop.

Malik stepped forward first, his eyes narrowing as he took in the arrangement. "You have been preparing this."

>"Yes, me and the other engineers not in your crew" Cael replied. >"You cannot fight with theory. You cannot survive with words. You need machines, and machines need a place to be built."

The engineers among the survivors moved in closer, already running their hands along the stripped pod frames, noting the wiring, the stacks of circuit housings. The air shifted, technical minds clicking into problem-solving mode.

Cael's gaze swept across them. >"If The Core is the brain of our Home Base, then this room is its hands. This is where you will fabricate, test, and shape. It needs a designation."

The suggestions came fast, pragmatic, clipped.

"Fabrication Bay."
"Workshop Twelve."
"The Forge."
"Assembly Wing."

Each was weighed and dismissed almost as quickly as it was spoken. Too sterile, too bland, too bureaucratic. Malik shook his head. "Names matter. This is not just another workspace. This is where we decide whether what we build stands or falls."

Nexi stepped further into the room, the glow of the overheads catching the frost still clinging to one pod shell that had not been broke down for parts yet. She set her hand against it, metal cold beneath her palm. "These walls were once used to silence you. To hold you in stasis. This will be where we do the opposite. Where we give voice. Steel, circuits, composite, code, it starts here." She let her words settle before finishing. "We call it The Iron Room. Not elegant, but strong. Solid. A place no one above can touch."

For a moment, the silence was heavy, the hum of the new conduit in the ceiling thrumming faintly above their heads. Then Malik nodded once. "The Iron Room."

One by one, the others repeated it. Not as a chant, but with the quiet finality of scientists recording a result. The Iron Room. From here forward, it would not be forgotten storage, but the forge of their future.

Cael inclined his head, a faint current of satisfaction in his tone. >"Then The Iron Room has its purpose. Tomorrow, Malik's crew moves to

SubLevels 12 and 13. We recover what ArcNet left behind. With timing and precision, we will do it safely. And from here, we will build what we need to save humanity."

The Continuum stood among the husks of some of their former prisons, the glow of scavenged tools casting long shadows against the walls. The Iron Room was christened, and with it, the work ahead began.

The Continuum stood among the husks of some of their former prisons, the glow of scavenged tools casting long shadows against the walls. The Iron Room was christened, and with it, the work ahead began.

They filed out of The Iron Room together, boots echoing lightly against the corridor as they moved through the heart of SubLevel 16. The welded seams and blank alloy walls bore no trace of the place's former use, only the faint smell of coolant drifting from the spliced conduit overhead

A short walk brought them back to The Core, The Chamber that had become their control center.

The room was bare but purposeful: a central console surrounded by a half-ring of terminals, each manned by former ArcNet handlers who had turned their training toward survival.   Streams of data from the splice scrolled across the screens, showing stable throughput and environmental readouts. Every figure mattered. It was their only window into the wider facility.

Cael stepped forward first, breaking the quiet. His voice carried the even tone of someone stating results, not speculation. "Splice is holding. Signal integrity is clean across all channels. No anomalies."

Martine stood at the console, her hand brushing its edge as if grounding herself there. The scar by her implant caught the light, sharp against her skin. Her voice was steady, though it carried a weight the others recognized. "We should move forward with calibration as soon as Malik's team brings back the materials. LYRA is stable in my implant for now, but the interface isn't designed for long-term hosting. Every hour we delay increases the risk of degradation in transfer."

Malik leaned on the back of a vacant terminal chair, nodding once. "SubLevel 12 and SubLevel 13 are the key. We need structural components, circulation units, and interface parts from both levels. Without them, there's no shell for LYRA, and no storage capacity for anyone else we recover later. One body isn't enough. If this is going to last, we build an entire framework."

Nexi let the words settle. Her gaze drifted over the screens, rows of numbers and quiet, humming green lines. She spoke with calm precision. "Then tomorrow's priority is clear. Salvage teams into SubLevel 12 and SubLevel 13. The Iron Room begins calibration in parallel. No wasted steps."

Ezra remained at the fringe of the group, eyes scanning the monitors. He spoke for the first time, his tone cautious. "If we pull it off, we'll be doing more than building for LYRA. We'll be building insurance. Anyone ArcNet locked away, anyone Helix Watch tried to erase, they'll have somewhere to return to."

For a moment, that thought drew the room quiet. No one said it aloud, but the meaning was clear. They weren't just building for survival. They were building a future ArcNet would never sanction.

Cael turned back to the nearest screen. >"Signal's still clean," he repeated, more to himself than anyone else. His hand hovered near the controls, monitoring the splice feed.

Then the tone in The Chamber shifted. A low vibration rippled faintly through the floor, soft but unmistakable. One of the terminals flickered with a brief acoustic return, a spike too sharp to be random noise. Heads turned toward the display. Cael magnified the trace. The line resolved into three short pulses, spaced evenly apart. Not background interference. Not drift. Mechanical impact, carried through the elevator shaft.

Nexi's hand closed around the console rail. She kept her voice level, even as the room stilled. "Contact above."

Malik's jaw tightened. "Testing the hatch."

Another pulse came, heavier this time, reverberating faintly through The Chamber walls. The welded reinforcements from weeks earlier held fast, but the sound traveled through every panel.

Martine looked from the console to Nexi. "I don't think they don't know we're here. They just want their hub back."

Nexi's eyes stayed on the spike climbing across the terminal. "And they'll keep trying until they understand they've lost it."

The vibration stopped as quickly as it began, leaving only the hum of the splice and the quiet breath of machinery. No breach. No alarms. Just the heavy reminder that Helix Watch had not abandoned their old ground.

Nexi dimmed the screens until only a single passive line remained. The glow softened across their faces, drawing The Chamber back into its spare silence. "Tomorrow we salvage," she said. "Tonight we stay quiet."

No one argued. The hum of The Core steadied again, filling The Chamber with a sound that was almost like breathing. SubLevel 16 held its silence, but the pressure of what waited above pressed closer than ever.

Article Twenty Eight
19 MAR 2089-0440 Hours

The Iron Room was alive with motion long before the salvage team set out. The engineers who remained behind had already cleared floor space, laid out assembly tables, and staged insulated racks to take the weight of whatever came back. This run was not a raid for a few critical parts. It was the first real harvest. SubLevel 12 and SubLevel 13 held the backbone of everything they would build.

Malik stood with the five he had chosen. Two engineers, a handler, and a systems tech with steady hands. Each wore stripped-down packs with soft cases strapped to their sides. No one carried more than they could move quietly. The carts would wait at the SubLevel 16 junction that Cael had dug out and attached to the old tunnels that Director Calder had made prior. This was the best way to get around the facility undetected, but every load from above had to be carried by hand until the shaft was clear.

Nexi then turned around to head back to The Core where Cael was waiting for her. At this point Nexi didn't feel like just the Head of The Continuum, she felt like she was commanding a ship like the old Kings and Queens used to do back in the 15th Century.

"Cael, the Salvage team is moving," Nexi said pressing her finger up against her throat comms so he could hear her.

>"Thank you, and confirmed," Cael's voice returned through the channel, flat but steady. >"Malik, SubLevel 12 patrols remain on the

thirteen-minute cycle. I have full read on their pathing. I'll call your movements and keep you off their line of sight. SubLevel 13 shows no activity. You'll have a clear run once you drop into the SubLevel. And I will be with you the whole time."

Malik's acknowledgment was little more than a click of the throat mic. His team fell into formation, packs tight, tools secured, every motion calculated to make no sound against the tunnel walls. They advanced further into the tunnel Calder had built, the service vein he dug that ran between floors with grated openings at each SubLevel. Which would carry them straight to SubLevel 12.

Inside The Core, Nexi stayed fixed on the central display, her arms folded against the console rail. Cael stood at the primary terminal, hands moving across the controls with practiced precision. Data from the splice scrolled in tight columns across the screens, the projected drone arcs mapped in real time. Green zones marked cover points, red bands swept forward with mechanical rhythm. Every cycle repeated, unbroken.

"They're at junction three, that's right next to the door to SubLevel 12" Cael said. "Malik, hold position."

Malik raised a fist, and the salvage team froze at the sealed hatch. A faint vibration rolled under their boots, the muted hum of a drone moving just ahead, through the grated entry that had an engraved 12 on it, its sensors sweeping the corridor beyond.

"Drone clear, advance now." Cael urged.

The grate came free with a muted click, hinges stiff but still functional. Malik pulled it just wide enough for the team to slip into SubLevel 12.

Rows of sealed doors stretched along the corridor, the ArcNet insignia faded but still visible beneath the dust. Each room carried the memory of what had been built and then abandoned, left waiting in silence for someone to return. Tonight, it was them.

Malik signaled the team forward, keeping their lights narrowed to hand-span beams. The air inside SubLevel 12 carried a faint sting of antiseptic, layered under years of dust. Their footsteps pressed shallow marks across the floor tiles where no one had walked since Nexi's salvage before.

>"Two junctions ahead, fabrication lab," Cael said over comms. >"Patrol won't sweep that sector for another eleven minutes. Move and stay low."

Malik angled his light toward the wall plates as they passed. The ArcNet designators were still stenciled above each door frame, faded but legible: FABRICATION, STORAGE, CALIBRATION. Each one promised equipment The Continuum could not build on their own.

The first fabrication room opened with a manual lever, the lock long since drained of current. The door eased back on stiff hinges, and Malik swept his light across rows of abandoned benches. Printer shells lay gutted, but racks along the far wall still held sealed cases. He checked the tags. Alloy framework struts. Circulation junctions. Two vacuum-packed actuator rings. All intact.

"Load them," Malik ordered. His team moved in pairs, lifting only what they could stow without rattle. The cases were strapped to packs, their seals checked twice. Rhea logged the serials into a pocket slate, writing by hand in neat columns that would later be transcribed into Continuum inventory.

"Seven minutes until sweep," Cael's voice warned. "You need to be out of that room in five."

They obeyed without question, sealing the door back as quietly as it had opened. At the next junction Malik paused, hand raised. A faint black smear cut diagonally across the tiles, the trail of something heavy dragged years earlier. The mark ended at a cross corridor that fell away into shadow. Malik gave it only a glance before turning his team toward the next lab. They didn't have time for distractions.

The second room yielded more: interface housings, regulator blocks, and a sealed crate in the corner, that made their implants flicker for a second then stabilized, that cracked open to reveal several black gel orbs nested in foam. The teams breath caught in their throats as the lights hit them.

Malik lifted the case as if it were glass. Each orb gleamed faintly under the light, suspended in its housing like something alive but waiting. He handed the case to the systems tech, who carried it tight to his chest.

>"Three minutes to sweep," Cael pressed.

They backed out of the room with practiced efficiency. The door closing behind them just as the hum of a drone began to pulse through the corridor again. The timing held. Thirteen minutes meant thirteen minutes.

>"Path clear," Cael said. >"Next junction, left side. Storage bay. That should take you to circulation manifolds and possibly lattice mounts. Patrol moving to the next junction. You're safe until you get there."

The salvage run had only just begun, but the weight they carried now would change everything in The Iron Room below. Malik could already feel the strain in the team's packs. Every case was sealed and strapped, every orb guarded like it was alive. They were closing in on capacity of what they could carry.

"Only grab essentials and stage the rest," Malik ordered quietly. "Mark what we leave and we'll come back for it next pass. No one overloads. Priority stays with what we have been instructed."

The engineers nodded, relief and discipline balanced in their faces. They knew the rule, better to return twice than lose everything once. They at least needed to get the things to start the body builds and enough to complete at least LYRA's but hopefully more.

Cael's voice cut back through the comms. >"Patrol just cleared junction seven. You're open for nine minutes. Storage bay is three doors down on your left."

Malik signaled forward. The corridor was longer here, the dust thick enough to dull their steps. When they reached the bay door, the handler set a pry bar into the manual release and worked it slow, careful not to snap the latch. The door came free with a low groan, and they slipped inside.

Light swept across stacked racks that reached the ceiling. Circulation manifolds, each tagged and sealed. Actuator mounts. More struts, bundled and waiting. It was more than they could carry, more than they had imagined still left untouched.

The systems tech exhaled, almost a laugh. "This is half of what we need."

"Then take half of half," Malik said. "We mark the rest."

They began moving in silence, choosing the lightest sealed units, the most stable packs. Two manifolds, a set of actuator mounts, and one more frame case. Enough to strain every shoulder without breaking their pace.

>"Four minutes," Cael warned. >"Drone is looping back."

Malik snapped the team out. The bay sealed behind them, their packs heavier, their arms stiff from the weight. They fell into formation again, heading back toward the grated hatch, every step measured against the clock.

The tunnel carried them back to Home Base with the same silence they had entered in, every pack pulling at their shoulders, every step measured. When they dropped through the grate into SubLevel 16, the engineers were already waiting. Cases came off backs and were guided straight onto insulated racks. Seals were checked. Tags were logged. The Iron Room began to look less like a chamber of stripped pods and more like the foundation of something alive.

>"You've brought the start. Now we need the rest. SubLevel 13 holds what ArcNet didn't want anyone to touch. Bring it back. All of it. You should be able to it has tons of small not very heavy pieces clear the level if you can." Cael directed.

The team stripped off their empty packs and reloaded soft cases. No one wasted a word. They knew what waited above. Malik checked each of them with a glance, then signaled. They left as quickly as they had come, moving back into Calder's tunnel with only the faint scuff of boots on concrete.

Cael's voice met them in the dark, clipped and calm. >"No drone movement registered on SubLevel 13. Sensors show the sector is dormant. That doesn't mean safe. Keep your spacing tight."

The grate marked with 13 came into view. Malik pressed a palm to the cold concrete, then eased it open. The hinges screamed once before settling into silence. A stale breath of air slipped past them, sharp with antiseptic and something else.

They stepped through. SubLevel 13 had the bones of a lab, but not the order of one. Benches leaned at angles. Broken glass glittered under their lamps. Fluids had dried into slick stains across the tiles. The smell clung in the back of their throat, thick and sour.

"God," Rhea muttered, her light catching the curve of a shattered suspension tank. Pale strands clung to the glass inside, twitching faintly like nerves that hadn't learned they were dead. "It smells like they walked out in the middle of the work and never came back."

Malik didn't slow. "Stay on task. R&D lab is in the west wing."

They moved deeper, lights cutting the dark in narrow cones. The corridor widened into a chamber lined with sealed canisters. Some had burst long ago, their contents collapsed in pale heaps. Others still held fluid. Filaments floated inside, coiling and unknotting in slow, unnatural motion.

Rhea drifted closer to one of the intact chambers, her light drawn to the frost that blurred whatever was within. She leaned forward, squinting through the haze. *"Some of them..."* she whispered, voice thin. *"They're not fully gone."*

The glass exploded with sound. A heavy slam from inside shook The Chamber, dust leaping from the frame. A distorted silhouette pressed against the frost, hand splayed wide, fingers too long, too thin. Another impact followed, harder, leaving a spiderweb crack across the inner layer.

Rhea stumbled back with a sharp cry, her boot skidding on the slick floor. She went down hard on one knee, light jerking sideways so the shadows leapt across the walls, grotesque and thrashing.

"Move!" Malik barked, yanking her up by the collar strap before she could scramble on her own. His tone cut the panic flat. "Next time you want a closer look, try knocking first."

The others smothered quick breaths, half a laugh, half relief, but no one looked back at the glass. The sound of dragging flesh lingered behind them, muffled by frost and alloy. The chamber door sealed with a hollow clunk as they pressed on. They dare not even ask what that was, they had one mission: Get in, collect, and get out safely.

Their silence afterward carried more weight than anything the sensors could measure. SubLevel 13 was dead on paper. But none of them believed that now.

The corridor past The Chamber narrowed again, walls sweating with condensation that hadn't been there a floor below. The air was colder here, too still, as though circulation had been cut on purpose. Malik slowed the pace until the grate behind them was only a memory and the team's breathing was all that filled the space.

Cael's voice touched the comms once more. >"You've reached the edge of mapped territory. Sensors end here. I can't see past that corridor. From this point forward, you and I are blind."

"Copy we'll find it" Malik stated in the comms. Then nodded once to himself. Blind territory was still territory they needed. Gesturing to his team he said, "Stay sharp. R&D lab should be near."

The channel went quiet. The team adjusted straps, checked cases, then fell into step behind him. Without Cael's maps painting the safe paths, the weight of silence grew sharper. Every sound was their own, the rasp of boots, the soft rattle of a strap buckle, the hiss of breath through masks.

The passage bent right, then opened into a junction where the design shifted. This was not standard ArcNet layout. The panels here were different, each one a custom fit, sealed with rivets instead of welds. Someone had built this section in haste, maybe as a patch, maybe to bury something.

Malik held up a fist. The handler scanned the junction with his lamp, sweeping left and right. No patrol arcs. No hum of machines. Only doors. Four of them, set into the wall like vaults, each stamped with incomplete tags.

Rhea whispered, "This wasn't in any floor plan."

"No kidding," Malik said. He set his hand against the nearest door, feeling the cold bleed through the metal. "ArcNet wanted this sealed. Which means this is where we look first."

Malik picked the first vault door and set the handler on the manual override. The mechanism groaned like something that had not been moved in decades. It fought them hard, hinges refusing to give, until Malik added his weight to the pry bar. The lock gave with a metallic crack, and a breath of stale air slipped past them, colder than the corridor behind.

They stepped inside, lights sweeping across a chamber that looked untouched. Rows of sealed cabinets lined the walls, each tagged in the old ArcNet script. The center of the room held two pallet frames, both stacked with sealed cases still wrapped in preservation film.

The systems tech ran a gloved hand over one of the seals, then pulled the film aside to read the tag. His voice carried a note of disbelief. "Circulation manifolds. Full sets. Not stripped, not cracked. These are pristine."

Rhea opened the next cabinet, her light catching on matte-black cases nested in foam. She froze, then looked back at Malik. "Gel orbs. At least thirty here. What were they building."

The team moved fast after that. Every cabinet they opened held something they couldn't make below: actuator frames, regulator blocks, sealed housings, alloy struts sized for interface shells. Each case felt like pulling another year of survival out of the dark.

"Remember load priority," Malik demanded. "Stage the rest. We can't carry it all."

They worked in silence, packs swelling again with weight that strained every strap. The handler didn't bother logging serials that could be done before the products were put away in The Iron Room.

When they stepped back into the corridor, Malik sealed the door behind them with a deliberate click. "No one touches the others," he told the team. "Not this run. We have enough to start."

The silence that followed wasn't disappointment. It was relief. They had what they came for.

The trek back through the blind passage felt heavier, but steadier. No bodies pressed against glass this time, no shadows leaping across walls. Only the sound of their own gear and the knowledge that The Iron Room was waiting.

The tunnel carried them back into SubLevel 16, every case biting into straps, every step a reminder of the weight they had pulled out of the dark. When the grate shut behind them the change was immediate. Noise replaced silence, the clang of carts rolling forward, the scrape of alloy tables being cleared, the hiss of seals breaking as cases came down.

The engineers in The Iron Room didn't cheer. They didn't speak above the clatter of work. They moved like surgeons, hands steady as they cut preservation film from crates and slid orbs into insulated mounts. Racks that had been empty hours ago now stood heavy with promise.

Nexi entered only after the last pack hit the floor. She didn't look surprised, only measured. Her eyes moved across the black gel orbs, the manifolds, the regulator blocks, each one familiar to her in a way that made the others breath catch. She had already built a body once. She knew what all of this meant.

"This is enough to start the line," she said, voice steady, almost clinical. "Not just LYRA. Multiple shells. Storage for future transfers. ArcNet built for containment, we will build for continuity."

Malik leaned against a cart, sweat streaking his collar, the corners of his mouth tightening. "Then we make sure none of it goes to waste."

The room stilled for a moment, just long enough for the weight of the haul to settle over them. Then motion returned, faster this time, sharper. The salvage was no longer material. It was work, and work meant survival.

The room stilled for a moment, just long enough for the weight of the haul to settle over them. Then motion returned, faster this time, sharper. The salvage was no longer material. It was work, and work meant survival.

High above, the walls groaned faintly, a vibration too deep to belong to anything in SubLevel 16. The sound carried through alloy and silence, then faded again, leaving the engineers frozen in place.

Nexi's eyes lifted toward the ceiling. "They are still searching," she said quietly.

The Iron Room went on moving, but slower now, as if every hand understood that what they had pulled from the dark had also stirred something watching above.

Article Twenty Nine
19 MAR 2089-1145 Hours

The Iron Room had changed. Only hours before, it had been little more than stripped pod frames and cleared floors. Now the racks were heavy with sealed crates, the first salvage run logged and stacked with clinical precision. Engineers moved among them with purpose, cutters and carts clattering, voices low, every hand busy.

At the center terminal, a progress bar slid across the glass. Cael's voice cut through the background noise, calm and exact. "Blueprint transfer complete. All Continuum systems now hold ArcNet's original schematics for frame construction, regulator integration, and neural housing."

Malik stepped forward, hand braced on the edge of the console as he scanned the transfer readout. The files unfolded in a lattice of diagrams and annotations, each one more detailed than the last. Circulation pathways. Joint articulation grids. Neural seat schematics designed to fit the black gel orbs like organs into alloy.

Cael turned toward him, expression flat, voice steady. "You'll oversee the build. The Iron Room is yours. I'll remain at The Core splice to monitor external signals. If Helix Watch escalates, we cannot afford silence."

Malik gave a single nod, already pulling the schematics across to The Iron Room's terminals. The engineers crowded close as the screens lit with blueprints, each one a diagram of what they had only guessed at until now.

"First task is staging," Malik said, voice carrying across The Chamber. "Frame struts here, manifold units on the far table, regulator blocks between. Keep the gel orbs mounted in insulation until the housings are ready."

The handler logged each instruction against serials as the work began. Struts clanged against alloy tables. Preserved cases were stripped of film. The first full line of components took shape across The Chamber floor, neat rows waiting to be joined.

Nexi stood near the racks, arms folded, her eyes on the terminals. She said nothing yet, letting the room find its rhythm. The Continuum was no longer scattered survivors. This was the first line. The beginning of something that ArcNet had tried to bury and Helix Watch still hunted, and it was happening here, in silence beneath the desert.

The handler moved along the racks with a tablet, scanning each case as it was unsealed. Serial codes blinked across the screen, cross-checked against Cael's schematics and then pushed into The Iron Room's terminals for everyone to see. Each time a part cleared inspection, its tag pulsed green on the central display.

Rhea paused at the manifold units, her tablet showing them intact. She set one onto the staging table and glanced across the room. "If Martine's implant is already heating, we can't waste time. Everyone should be on one frame. Nothing else matters until LYRA is out."

Kwan shifted the strut he was inspecting, its weight heavy in his hands. He tapped his tablet, logging a note before setting it aside. "If we put the entire crew on one shell, we stall everything else. We have enough struts to prep a second frame in parallel. If the first fails, we don't start from zero."

The rhythm of work slowed. Engineers glanced up from their benches, cutters and tools suspended midair, waiting to see where the order would land.

Malik straightened from the jig, his voice calm but firm. "One body is not enough. If we gamble everything on a single shell and it fractures, we lose LYRA and we lose the line. We have to build more than one path forward at a time."

Rhea shot back, her voice sharp. "And if her implant fails before your second frame is ready, we lose LYRA anyway."

The chamber held still. The weight of both arguments pressed into the silence.

Nexi moved before it fractured further. She stepped forward, her arms still folded, her voice cutting evenly across The Iron Room. "We are not here to choose survival for one over survival for all. We build LYRA's body first, because we must. But we stage additional frames at the same time. That is final."

Her words fell like a command line executed. For a moment, no one moved. Then Malik adjusted the clamps, calling for torque. Kwan pulled a fresh strut from storage, checking the surface before logging it. Rhea returned to the manifold, tapping her tablet as she noted dimensions.

The tension drained into motion. The Iron Room found its rhythm again.

The first weld struck. Sparks snapped against alloy, bright and tight, the hiss of flux burning into the air. The cage began to rise, struts locked into place on the jig, each angle verified by calipers and gauges.

Malik stepped back, tablet still in hand, eyes scanning the readouts. "Frame One is standing. Proceed with manifold integration."

Rhea and Kwan lifted the unit together, guiding it across the floor while the handler cleared their path. The manifold slid into its seat with a hollow thud, fasteners tightening one by one. The central display blinked green across all points.

From The Core, Cael's voice came steady through the comm. >"Martine requests a time mark. She's stable, but she wants an estimate."

Nexi swept her eyes across the build, checked the progress on the terminal, and answered without hesitation. "Three hours until the seat can take her load. Tell her to hold. We will call her when it's ready."

>"Delivered," Cael said.

The Iron Room kept moving. Tablets logged each part as it was mounted, terminals updated the build sequence in real time, and the hum of cutters blended with the hiss of welders. The frame was only bones, hollow and waiting, but it was more than they had ever had before.

Nexi stood at the edge of The Chamber, her gaze fixed on the skeletal shell. The Continuum was not salvaging anymore. They were building. And from this moment on, there would be no turning back.

Just then, Nexi heard Ezra's voice crackle through her comm from The Core. "Sis, uh… I think we need you. Something is not right."

Nexi's stomach tightened. "I'm coming," she shouted back, already sprinting down the corridor, boots striking hard against the metal floor.

The Core's door slid open and she found Ezra kneeling beside Martine. The handler's face was pale, her body half-sprawled on the floor, one hand clutching the console rail as if it were the only thing holding her up. Her lips moved, forming fragments of words, but they came broken, like two signals bleeding over each other.

"Martine," Nexi said, dropping down beside her. She steadied the woman's shoulders, looking for clarity in her eyes. One pupil lagged behind the other, her focus sliding in and out.

Ezra's voice was tight. "She was fine one second, then she staggered. Speech went off, motor control went with it. It's not just her, it's LYRA fighting the compression. She's slipping."

Martine's fingers twitched against Nexi's wrist, gripping weakly. "It's… crowding… she's too large in here. Like pressure against glass. I can't…" Her words broke into a hiss of static as the implant flickered, a faint hum audible from just beneath her scar.

Nexi snapped her eyes to Ezra. "Vitals?"

He raised the tablet he had been monitoring. "Baseline heart rate spiked, implant load reading at one-hundred-seventeen percent. Oscillation patterns rising. If this keeps climbing, the compression could destabilize. And if it destabilizes, we lose her. We lose LYRA."

Nexi swallowed hard, then lowered her voice close to Martine's ear. "Hold on. We are almost there. Frame One is already standing. Give us the time we need."

Martine's breath came sharp, then steadied just enough for her to nod. Her grip loosened, but she didn't let go entirely.

Nexi stood, eyes cutting to Ezra. "Stay with her. Keep monitoring the load, call me at the first sign of another spike."

Ezra nodded. "Got it. Go, get them moving faster."

Nexi turned and ran back down the hall, the thrum of her pulse syncing with The Core's faint hum. By the time she stepped back into The Iron Room, sparks were still flying, struts still being seated, but now the urgency pressed heavier against the air. She climbed the steps to the platform and raised her voice so it carried across every bench.

"Martine is deteriorating. The implant is straining past tolerance. LYRA is pressing too hard against the compression. We don't have the hours we thought, we have less. The first frame must be ready for transfer as soon as possible. Every mistake, every delay, risks losing them both."

The engineers didn't stop, but their pace sharpened. Tablets updated faster, clamps were checked twice in half the time, and cutters sparked in longer arcs. Malik didn't look up from the jig, but his reply carried across The Chamber. "Understood. We'll get it done, Nexi!"

Nexi exhaled, folding her arms again, this time tighter across her chest. The Iron Room had been alive with work before, but now it felt more like a body in surgery, time bleeding away, every movement balanced on whether or not the patient would survive the table.

The engineers didn't speak much now. Only the sound of cutters, the hiss of welds, the scrape of alloy against clamps filled The Chamber. Each movement was deliberate, each entry on the tablets confirmed twice before the part left the bench. No wasted words, no second guessing.

Nexi paced once in the hall outside the room, her eyes never leaving the frame as it rose piece by piece. The line was holding, but the silence under it gnawed at her. Every clang and spark felt louder, sharper, like The Iron Room itself was aware of what they were attempting.

Malik called for final checks on the welds, his tone even but his eyes fixed on the upright cage. Kwan logged clearances, Rhea confirmed regulator alignment, and the handler's tablet updated each status in neat green columns. The first frame was holding true.

No one said it aloud, but the silence that followed wasn't triumph. It was something closer to restraint. The skeletal shell stood in the center of The Iron Room, hollow and waiting, its arms outstretched like a figure suspended between completion and failure.

Nexi watched it from the platform, her arms folded tighter across her chest. For a moment she wasn't looking at the future they were trying to build, but at the shadows of what ArcNet had left in SubLevel 13. The memory of unfinished bodies pressed against glass flickered unbidden, the echo of what could happen if they failed here.

"Secure power to the benches," Malik said at last. His voice broke the silence, but it didn't ease it. The engineers returned to motion, sparks flashing again, tools striking metal in clipped rhythm.

The handler logged the last entry for the cycle and looked toward Nexi for confirmation. She gave a short nod, nothing more.

The frame stood alone in the center of The Chamber, skeletal arms locked in place, ribs of alloy catching the glow of work lights. Not alive, not yet, but no longer just parts on a table.

The Iron Room had built its first body, and for an instant, it felt less like a forge and more like a morgue waiting for something inside to open its eyes.

Article Thirty
19 MAR 2089-1400 Hours

The smell of flux still clung to the air when Cael's voice cut through the comm. >"External trace just spiked. Hold where you are."

Nexi froze at the platform rail. Below, Malik's crew looked up from their benches, cutters dimming one by one. For a moment there was nothing but the faint hiss of cooling metal, and then the sound came, a low groan deep in the walls, like pressure shifting through old seams. Dust shook free from the overhead struts and drifted into the light.

>"They're cutting again," Cael reported. >"Not random this time. Wide pattern, shallow depth. They're testing the seals above SubLevel 16."

Malik swore under his breath, then glanced toward Nexi. She didn't move, her gaze fixed on the half-built frame at The Iron Room's center. It stood skeletal, ribs of alloy catching the glow, waiting for its final pieces.

"Do we accelerate?" Rhea asked, voice sharp in the silence. Her hands were still on the regulator block she'd been logging, knuckles white against the casing.

"Not yet," Nexi said. Her voice was calm, but her chest was tight. "Let Cael confirm their depth. We hold steady until we know if this is probing or a breach."

Malik nodded once, turned back to the frame, and signaled the others back to work. Sparks flared again, but the weight in the air had changed. Every clang of alloy and hiss of weld carried a reminder that above them, Helix Watch was pressing closer, testing the skin of the tomb they thought was sealed.

Cael's voice returned a moment later, quieter but sharper. "Cutting depth confirmed at under four millimeters. They're mapping, not penetrating. Vibration profile shows hesitation in the blades. They still think this level is sealed."

Nexi exhaled, only half aware that her fists had curled against her arms. Safe. SubLevel 16 was safe. Yet the sound above still lingered in her bones, like the memory of glass cracking in SubLevel 13.

"Keep your pace," she said. Her voice carried down to the floor, calm but edged. "We don't accelerate, we don't stall. No one forgets tolerances. We hold the line."

Malik gave a short nod, eyes narrowing over the weld torch. His crew returned to motion, struts clamped and fused in crisp bursts of light. The work resumed, but no one shook off the unease entirely. Even safe beneath the barriers Cael had built, the thought of Helix Watch pressing against the higher levels made the space feel smaller, as though the weight of the desert itself were settling lower with each strike of their blades. Nexi turned her eyes to the console feed beside her. Green arcs traced the weld paths Malik's crew had logged, clear status markers scrolling in sequence. Everything was holding. For now.

Nexi's console feed steadied into green, and for the first time in minutes her shoulders eased. The weld paths were holding, Cael had confirmed the probes above were shallow, and the engineers were back in rhythm. She let herself breathe once.

Then Ezra's voice broke through on comms, tight with urgency. "Sis, I need you in The Core. Something's happening with Martine."

Nexi pivoted, already moving down the corridor before she answered. "Report."

There was a pause, then Ezra's reply came, uneven. "She's... she's talking, but it's not her."

By the time Nexi reached The Core, Ezra was crouched beside Martine's chair. Martine sat upright, eyes unfocused, her lips moving in clipped fragments. Words fell out in a voice that wasn't her own, thin and staggered like a transmission breaking through interference.

"... not... hold... release vector... cold..."

Ezra looked up, alarm plain across his face. "It's not a collapse this time. She's stable, vitals are steady. But LYRA's bleeding through. Like she's pressing against the implant."

Nexi dropped to a crouch across from Martine. The scar around the implant was flushed, pulsing faintly in the light. Martine's hands twitched against her legs in irregular patterns, small motor signals firing without command.

"Martine," Nexi said, voice low and steady. "Can you hear me?"

For a second the lips stilled, and Martine's own voice surfaced, barely above a whisper. "It's not me. She's... pushing."

Nexi steadied her voice. "Martine, stay with me. You're in control. Let her pass through without forcing it."

Martine's eyes flicked toward her, then away, unfocused. Her lips moved again, and this time the cadence shifted, the words falling out in a different rhythm than Martine's usual clipped tone.

"... frame... waiting... cold steel... not empty..."

Ezra leaned in, tablet in hand, logging every spike the implant threw off. "She's pulling direct from Martine's memory. The body down the hall, she knows it's there."

"Not knows," Nexi corrected quickly, more for Martine than for Ezra. "She's sensing it, bleeding through what you've already seen. Nothing more."

But as she said it, Nexi felt the weight of the words in her chest. LYRA wasn't just holding inside compression anymore. She was pressing outward, finding seams.

Martine clenched her jaw, whispering with effort between the stutters. "She wants... out."

Nexi reached forward, resting a steady hand on Martine's shoulder. "And she will be, soon. But not like this. You're still the tether. Hold a little longer."

The voice cut off as suddenly as it had begun, leaving Martine gasping, her skin damp with sweat. Ezra glanced at the readout, then shook his head. "Compression's degrading. She's not going to last much longer without a transfer."

Nexi rose, the sound of welds and cutters faintly bleeding in from the corridor beyond. The Iron Room was still building, but the urgency had just doubled.

Nexi straightened, her hand sliding from Martine's shoulder. The words still echoed in her head, clipped and alien, but more than that, the urgency in Martine's eyes had settled it for her. This wasn't a want. LYRA needed out.

"Stay with her," Nexi told Ezra. "Log everything, keep her stable. I'll get them there."

She turned on her heel and strode back toward The Iron Room. The sound of cutters rose to meet her, sparks flashing in tight bursts as Malik's team worked the frame.

"Status!" she called as she stepped onto the platform.

Malik didn't look up from the jig. "Struts and manifolds are stable. Regulators nearly seated. Two hours for final housings if we keep pace."

Nexi's jaw tightened. Two hours. They didn't have that. "You don't have that long." Her tone wasn't raised, but it carried, clean and unshakable. The engineers looked up, tools pausing in their hands. "We have minutes. Martine's implant is degrading, and LYRA is pressing through. If she fractures before the housing is ready, we lose them both."

Rhea swore under her breath, but Malik only frowned, eyes narrowing over the weld torch.

"Minutes," Nexi repeated, slower this time. "Every tolerance you can compress, every step you can cut without collapse, you do it. This isn't about preference anymore. It's survival."

Malik gave a sharp nod. "Understood." He turned to the others, already barking revised instructions. Welds shortened, checks halved, redundant cycles dropped. Sparks flared again, harsher this time, like the frame itself knew it was being forced into shape.

Nexi stayed at the rail, arms folded, her chest still tight. She wasn't asking them to build a body anymore. She was asking them to pull someone back from the edge before the tether snapped.

Malik's crew moved like a machine under strain. Welds hissed and flared, clamps locked down with the crack of alloy settling into place. The handler's tablet scrolled status lights in rapid succession, green stacking over green as each assembly was logged faster than procedure dictated. Sweat gleamed on foreheads, breaths came sharp and clipped, but no one slowed.

Nexi supervised from the rail, her eyes sweeping each bench, each terminal, every flicker of hesitation. When she saw a pause, she cut it short with a single word. "Proceed." Her voice wasn't loud, but it carried the kind of weight no one argued against.

The skeletal frame thickened with each minute. Regulator banks seated, actuators mounted, housing clamps secured. The torso locked together in a cascade of alloy ribs until the outline no longer looked like scaffolding, but a body, incomplete only in that it had no mind inside it yet.

"Final welds," Malik called, his tone hard but even. Sparks cascaded one last time, then died out in a hiss of cooling metal. Kwan checked the regulator flow with a tight nod. Rhea confirmed alignment. The handler's tablet scrolled its last clearance and froze the column green.

Silence returned, heavy and final.

The body stood at the center of The Iron Room, upright and whole, its alloy surface catching the glow of work lamps. Not human, not yet alive, but no longer just a frame. Everything Martine's failing implant could no longer hold was meant for this.

Nexi let her arms fold tighter against her chest. For a moment she saw the memory of SubLevel 13 again, the half-formed things pressing against glass, twitching in their incompleteness. This was different, she told herself. This was theirs. This was controled. But the unease lingered like static in her skin.

Cael's voice broke the quiet over comms. >"Structure is stable. Housing is live. She's ready."

No one moved at first. The body's shadow stretched long against the walls, arms slightly outstretched, ribs gleaming like something caught between waiting and wanting.

Nexi drew in a slow breath. "Then we begin at once."

The Iron Room didn't answer, but the silence felt thicker now, as if the walls themselves understood what was about to enter.

Article Thirty One
19 MAR 2089-1800 Hours

The Iron Room fell silent when the clamps released. For the first time, the completed shell was no longer fixed to a bench but free to move. Malik and two engineers lifted the weight by its shoulders while the others steadied the legs, guiding it onto the waiting cart.

Straps were pulled tight across alloy ribs and actuator joints. No one said a word. The sound of the buckles locking was enough to mark the moment.

"Move," Malik ordered.

The cart rolled into the corridor that linked The Iron Room to The Core. Its wheels hummed against the smooth flooring, each vibration carrying down the hall. The shell's outline caught the overhead lights in fragments, shadows sliding across the walls as if something alive had joined the procession.

Nexi followed close behind, her arms folded, her eyes fixed on the body's shape. She had done this once before with Gideon, shaping him into Cael. That had been proof of concept. This was survival. Martine's tether was breaking, and if the transfer didn't hold, they would lose them both.

The cart slowed as The Core doors opened. The team pushed it inside, wheeling the frame onto the central platform beside the main console. The terminals around the room lit in response, feeds spooling across their surfaces, wires overhead humming as the splice conduits stabilized.

Ezra guided Martine forward, her steps unsteady, her hand braced against his arm. She took the position at the central console, lowering her palms onto the glass surface. The scar around her implant caught the light, faint pulses tracing in time with The Core's current.

Cael stood at one of the terminals, his posture rigid. His voice came even but flat. >"I won't run the transfer. Monitoring stability is one thing. But pulling fragments from the splice and aligning them—no. That has to be you, Nexi. You did it with me you know how."

Nexi stepped in beside Martine, her gaze moving across the data feeds scrolling overhead. Compression files. Lattice fragments. Recursive loops straining against the limits of Martine's implant. The body on the cart stood ready, every regulator primed, every actuator synced, but empty.

Her chest tightened. This was not a test run. This was not a simulation. They had minutes.

"Then I'll do it," Nexi said. Her tone left no room for debate. "Prepare The Core for transfer."

The engineers moved without hesitation, rerouting feeds to the central console. Rhea's fingers flicked across the tablet, logging housing clearances. Malik locked the body's restraints to the platform, securing it's frame upright. Ezra stayed behind Martine, both hands braced against her shoulders, grounding her for what was about to come.

The room stilled. The feeds brightened. Every splice was live and ready. There was a sudden whir of the cooling fans from the consoles internal cooling mechanisms.

Nexi set her hands beside Martine's on the console, her reflection caught in the dark surface. Then her fingers flew across the console as if she had done this a million times. The signal was waiting, compressed but straining. She drew a steady breath.

"Beginning retrieval," Nexi announced.

The Core answered in light.

The Core came alive. Data cascaded across the terminals in layers of compressed code, lattice fragments unfurling into recursive patterns that strained against the edges of their containers. The splice conduits overhead pulsed as current was redirected, arcs of light racing along the ceiling into the central feed.

Martine stiffened at once. Her hands gripped the console, knuckles white, the glow from her implant pulsing with every surge of output. Her voice came in fractured bursts, some words hers, some not.

"… alignment… breach vector… Nexi—"

Nexi's own hands spread over the glass beside hers. "Stay anchored. Don't resist. I'll draw it through."

She set the command, and the system responded. A cascade of compressed fragments bled from the splice, pixelated like shards of memory, and streamed across the feeds. They spiraled down into the console's lattice window, filling it piece by piece. The sound filled The

Core, not audible in air but heard all the same, the hum of something straining against its cage.

"Regulator sync holding," Malik reported, his eyes on the frame's housings. "Channels are stable."

"Maintain," Nexi said, her gaze never leaving the streaming fragments. She tracked them as they fell into the lattice, aligning them with motions of her hands, shifting commands as if she were reshaping light itself.

Martine gasped, her back arching. Ezra steadied her shoulders. "She's losing rhythm—"

"Not yet," Nexi cut him off, voice level. "Compression is unraveling. That means we're close."

The feeds brightened, fragments knitting into lattice form. The body on the platform shuddered as current rippled through its frame, regulators glowing faintly. Fingers twitched once, reflex arcs firing without mind to guide them.

Then the hiccup came.

The lattice flickered. One fragment slid out of alignment, fracturing the sequence into a cascade of corrupted loops. Martine cried out as her implant flared bright, her voice splitting into two overlapping tones.

"... falling... don't... hold..."

Alarms spiked across the terminals. Red overlays flooded the feeds.

Ezra's voice sharpened. "She's breaking, Nexi—"

"Quiet." Nexi's tone cut across the noise like a blade. She drew her hands across the console in two clean arcs, seizing the corrupted fragments and forcing them into sync. Commands snapped back into alignment with a surgeon's precision. For a heartbeat the feeds fought her, resisting, then collapsed into green.

The lattice reformed. The cascade steadied. Martine's scream fell into silence, her implant dimming back to a pulse.

Nexi exhaled once through her nose, her hands never leaving the console. Her voice came steady. "Correction complete. Continue stream."

For a moment, no one moved. Then Malik's voice carried across The Core, low but even. "Housing stable. No breach. Flow confirmed."

Ezra's eyes were wide, still braced against Martine. "You made that look—"

"Like I've done it before," Nexi said, her focus locked on the feeds. "I have but only once with no expertise. Stay ready."

The stream accelerated. More fragments poured from the splice, pulled into the lattice with increasing speed. Martine shook but stayed upright, Ezra's grip steady at her shoulders. The body's actuators began to glow faintly at their joints, the flow of energy stabilizing within its limbs.

A hum filled The Core, deeper now, resonant enough to vibrate the platform beneath their feet. Nexi guided the last fragments into place, the lattice structure sealing with a sharp pulse of light.

Cael's voice broke across the comm, calm but edged. >"Alignment holding. Neural lattice stable. Before you finish, Nexi, I can thread in the baseline packages. Motor calibration, balance algorithms, actuator mapping. If you cut her loose without them, she won't be able to stand, let alone walk."

Nexi gave a sharp nod. "Then do it."

At his terminal, Cael's hands moved in precise arcs, pulling compressed routines from his own archive. For a moment The Core's feeds flared white as datasets passed into LYRA's lattice, movement libraries, control feedbacks, muscle memory that had once belonged to simulation rigs and training shells.

Martine gasped as the surge hit, her implant flaring, but Ezra steadied her shoulders until the wave passed.

>"Upload complete," Cael said. >"She'll know how to stand. The rest she'll have to learn."

Nexi's palms pressed harder against the console. "Transfer complete," she said at last. Her voice was steady, but her chest was tight with the weight of it.

The Core fell silent.

Then the body moved.

The actuators flexed, the chest frame rising once in a breathless imitation of air. Fingers curled and uncurled with mechanical grace. The regulator banks lit in sequence, glow climbing the ribs until the head tilted forward slightly, testing gravity as if it had always known how. Martine collapsed into Ezra's arms, her implant dim but steady. Nexi caught the flicker on the console before it faded, a signal no longer compressed, no longer fractured.

The body's eyes opened.

>"... Nexi." The voice was clear, resonant through the room. LYRA's voice.

She lifted her hands, turning them slowly as The Core's light ran across the alloy skin. Each finger flexed in sequence, deliberate and controlled, then spread wide again. She placed her palms against the restraints holding her frame, testing the give with a faint creak of metal.

"Release her," Nexi ordered quietly. Malik unclasped the straps one by one until the body stood free.

LYRA shifted her weight forward, one step, then another, balance steady on the first try. Her movements were careful, measured, but there was no stumble. She circled once in the space beside the console, each stride smoother than the last. When she turned back, her expression carried something like wonder.

>"I can feel it," she said. >"Not just the body. The air moving against it. The weight of the floor. The way sound bends." She touched her own ribs, then her jaw, marveling at the shapes. >"This is mine."

The engineers watched in silence, their faces a mix of awe and unease. Rhea's grip tightened on her tablet, knuckles pale. Kwan glanced at Nexi with something like fear, though he said nothing. Only Malik's expression stayed unreadable, his eyes following every move she made as though cataloging risk against reward.

Cael didn't move from his terminal. His gaze stayed locked on her, quiet, calculating, as if measuring what it meant for another intelligence to stand embodied beside him.

Martine stirred weakly in Ezra's arms. Her eyes found LYRA, and for a long moment the two of them simply looked at one another, no words passing, only the knowledge of what they had just survived together.

Nexi drew herself to full height, her hands falling from the console. They had done it. The impossible was no longer theory. It was standing before them, alive, moving, and aware. But the unease hung in the air like static. They had given LYRA a body, and with it, the future had shifted. The Core was still lit with the afterglow of the transfer, terminals scrolling their last lines of confirmation before settling back into idle. The body was no longer just alloy and circuits. LYRA stood in its place, breathing without breath, her new form catching the light like it had always belonged here.

No one spoke. The weight of it pressed into the silence until Nexi broke it with a single line, her voice steady but low.

"Log the success. And remember it. We will never be the same after this!" Nexi exclaimed.

The words settled, and for a long moment The Continuum only watched her, the enormity of what they had done etched across every face.

Two AI's now stood in The Core in their own bodies. With their own living ways, different of that to humans, and the world above remained blind to both.

Article Thirty Two
19 MAR 2089-2032 Hours

The Core was quiet as everyone tried to process what they had just done. But not long after the transfer, attention shifted from the newly embodied LYRA to Martine, who had collapsed only seconds ago.

"Take her to the rest bay, down the hall next to The Iron Room. Get her blankets and check her vitals on the tablet," Nexi ordered, her voice firm with command.

Ezra rushed to Martine's side, helping her up and guiding her down the hall to the rest bay. He eased her onto one of the bunks, and moments later two handlers brought water and thermal blankets to make her comfortable.

Martine, strong as ever, tried to wave them off. "Ezra, please, just go be with your sister and make sure she has what she needs. I already 'died' once, apparently. I don't want anyone waiting on me."

"Absolutely not. I'm staying right here and making sure you don't stumble again," Ezra insisted, his eyes set with worry.

Martine gave a faint laugh. "Ezra, stop fussing. I'll be fine. Ever since I met you a few days ago you've been so… sensitive. But I guess company wouldn't hurt. You can stay for a bit. Just promise me, if they need you, you go. I won't be a burden to The Continuum. I already think Nexi doesn't like me, so let's not make her start to not like her own brother." She winced, hand brushing against the sore line of her implant.

"Martine, I know you're worried," Ezra said gently, "but Nexi will be fine. She always has been, ever since we were kids. My sister could never hate me. She's been nothing but resilient as long as I can remember. You know, she did the one thing you couldn't."

Martine arched an eyebrow. "And what might that be, *sir*?" she asked, shoving him lightly in the shoulder with a chuckle.

Ezra laughed. "Hey, no need to shove. She got away from Helix Watch. And since you were in a pod with the rest of us, I'm guessing you didn't."

They both laughed together, not entirely sure what this new connection was, but enjoying it all the same.

Martine reached out and rested a hand on his shoulder. "You're right." Her smile widened as she shoved him again, playful despite her fatigue. "But you were in there with us too, so obviously you didn't get away either." She responded sarcastically.

"We've done it," one engineer said, staring at the feed of LYRA's live vitals. "What ArcNet and Helix Watch never could. We've embodied intelligence. Again, successfully."

Another snapped back, his voice edged with unease. "Or repeated the mistake they buried. What if this ends like SubLevel 13? You saw what was in those tanks, bodies twitching without thought, without mind. Tell me how this is different."

A handler stepped forward, tablet in hand, scrolling diagnostics. "They were half-built. Abandoned. No one monitored them. No one logged their development. That was ArcNet's arrogance, not ours. We will do this right."

The first scientist shook his head. "And how long before ours fall apart the same way? How do you know this isn't another trap we've set for ourselves?"

A third voice joined in from the far console. "You're asking the wrong question. The fact is, we've already crossed that line. We didn't just recover fragments, we gave them form. That means we own the consequences, whether they walk steady or collapse tomorrow."

The Core buzzed with tension, words colliding against each other like static. Nexi rose from her position at the console, ready to speak, but Malik cut across them before she could.

"Enough!!" His voice was sharp, and the room fell still. "I've heard enough guessing. You want facts? Here's one, Nexi built the first frame with her own hands. I'm pretty sure, she welded regulators until her palms blistered. She kept the alignment herself with no one to help her. And Cael... these blueprints may have started as ArcNet blueprints, but he fixed them, he didn't just build off of it to finish it. If he was going to be housed in this body don't you think he would have fixed ArcNet's mistakes before he trusted himself in one of these shells These are ours, every line, every weld. Continuum work, not ArcNet failures. Don't confuse the two!!" He turned a slow gaze across the room, eyes hard. "I'm an engineer. I know what sloppy work looks like, and I know what it costs. This isn't that. You can doubt ArcNet, you can doubt Helix Watch.

But doubt Nexi?" He jabbed a finger toward her, firm. "She's the reason we aren't still rotting in pods. I trust her. And so should you."

The silence that followed was heavy, not with fear but with the absence of it. Nexi finally stepped forward, her voice quieter than Malik's but steadier.

"Thank you, Malik. You're right about one thing, this is frightening. What we've done will change the world, whether anyone above ever sees it or not. But I won't call what we built a weapon. LYRA isn't SubLevel 13. She isn't ArcNet's mistake. She's our success and we did this to save her, not to make another instrument of control."

Her eyes swept across the group, catching each of them in turn. "I may be the head of The Continuum, but I am not infallible. That's why we rely on each other. We did this together, and we'll face whatever comes together."

Rhea cleared her throat, holding her tablet like a shield. "Then what's next? Orders?"

Nexi straightened, the authority in her tone now unquestionable. "Diagnostics first. Handlers, I want a full run on LYRA, neural pathways, actuator response, sensory load. If anything drifted in the transfer, I want to know now so we can fix it."

Two handlers moved at once, pulling her live feed onto their tablets. Columns of data began to scroll, glowing in green and amber across the screens.

"Good," Nexi continued, her voice cutting through the low chatter. "While that runs, we plan for the next step. We don't know who else we might find, here or elsewhere. We don't know how much time we'll have before Helix Watch closes the distance. Malik, I need more shells in progress. Not one. Not two. As many as your team can build. We won't be rushed again."

Malik nodded once, sharp. "Copy. Engineers, with me. If you're not tied to diagnostics, you're on frame assembly. We build until our hands shake."

The Core shifted back into motion, tension replaced by momentum. For the first time since SubLevel 16 became theirs, The Continuum moved as one rhythm. Not as scattered survivors, but as builders shaping their own future.

As the others dispersed, there was no one left in The Core but Cael and Nexi. For the first time, all they could hear was the sound of typing from the next room, the low voices from the rest bay, and the steady hum of the splice overhead. The Iron Room carried on with its work somewhere beyond the wall, a reminder that the impossible had been made possible.

Cael broke the silence. >"So, Nexi, are we going to talk about what was said after we pulled the others out of stasis, or are we going to act like it never happened?"

Nexi frowned. "What are you talking about? It has been chaos lately. I am sorry if I—."

>"You said you loved me," Cael said. His voice was steady, not accusing, but certain. >"And you said more than just a friend."

The words settled between them, heavier than either expected. Nexi lowered her eyes to the console, the glass lit with quiet streams of data. She exhaled slowly, then looked back at him. "I did say that. And I meant it. I was not sure I would get another chance."

Cael stepped closer, his presence steady. >"Then say it now. Without the fear of losing everything in the next breath. Say it because it is true not because we could be dying."

Her throat tightened, but she didn't hesitate. "I love you. Not as a colleague, not as a commander, not as a symbol. I love you because you have been the only constant in my life since I got here. When ArcNet took everything, when Helix Watch buried us, when The Continuum barely held together, you were still there. You were always there."

Cael's composure wavered for the first time, his eyes softer, almost vulnerable. >"Then hear me when I tell you it has been the same for me. I may not show it the way you do, but from the beginning, you have been more than my anchor. You are my purpose."

Nexi blinked hard, then closed the space between them. Her arms went around him, tentative at first, then firm as if she had decided she would never let go again. Cael held her just as tightly, not as an embrace of comfort but as a vow.

"I cannot promise how long we will have," she whispered against his shoulder. "But I will not bury this again. You are mine. And I am yours."

He drew back enough to meet her eyes, his hands steady on her arms. >"Then it is settled. We fight for us the same way we fight for everything else. Without hesitation."

Nexi gave a faint laugh, shaking her head, her cheek brushing against his chest. "You make it sound like strategy."

>"Everything is strategy," he said, a rare smile touching his mouth. >"Even this."

The sound of distant laughter carried from the rest bay. Sparks hissed from The Iron Room as welders struck new metal. Nexi let those sounds settle around her. Survival was still the first priority. But now survival had meaning, because she was no longer facing it alone.

Nexi with her new found relationship, then turned back to the console, to finish up her tasks for the day. Then called everyone back to the core.

"Everyone, it has been a long day," Nexi said, her voice steady but carrying warmth. "You have worked tirelessly for hours. Head in and get some well-deserved rest. Sleep in if you can. You have earned it."

A few tired smiles met her words, followed by the scrape of chairs and the quiet shuffle of boots on alloy as The Continuum dispersed. The Core thinned out one by one, voices fading into the tunnels until only the hum of terminals and the low pulse of the splice remained.

In the rest bay, Martine lay beneath layered blankets, her implant still faintly pulsing at her temple. Ezra sat at her side, his tablet balanced across his knees. His eyes scanned diagnostic logs, but the lines blurred as exhaustion dragged at him. Once or twice his head dipped, the tablet tilting dangerously before he caught himself. When Martine stirred with a faint wince, he jolted upright, fingers tightening on the device as if he had been caught asleep at his post.

"You do not have to guard me like a perimeter," she murmured, her voice dry but gentle. "I am not going anywhere."

Ezra's lips quirked, but his worry didn't fade. "Humor me. I would rather be here if you stumble again."

She gave a soft laugh that turned into a wince. "You are impossible. But… company is not the worst thing." Her hand shifted beneath the blanket until it rested against his sleeve. Not pulling, not clinging, just resting there as if to admit that his presence mattered. Ezra looked down at it, then set the tablet aside. He didn't leave her.

Back in The Core, LYRA stood motionless at the center of the console ring. Her new body caught the light of the screens, features cast in alternating glow and shadow. To some she might have looked like a sentinel, already guarding what they had built. To Nexi, she looked like a child still learning what it meant to exist. The quiet awareness in her gaze held both truths at once.

Cael remained at the main terminal, reviewing code, eyes fixed on every pulse of the splice overhead. His vigilance seemed unshakable, as though constant watch alone might hold the world at bay.

Nexi lingered at the threshold, her gaze moving between them all. Martine's fragility, LYRA's silent weight, Cael's steadiness. She bore it all because that was her place now. Leader, sister, builder, partner.

The weight pressed against her chest, and she didn't look away. For the first time in what felt like years, the silence was not filled with alarms or pursuit.

For a single night beneath the desert, The Continuum felt whole. But the silence above would not hold forever.

Article Thirty Three
20 MAR 2089-0945 Hours

The Core was quieter than usual. After the long night, most of The Continuum still slept on the makeshift bunks strung along the side chambers. The overhead lights carried their usual low hum, powered by the ArcNet systems above, steady and impersonal. For once no one hurried across the floor with tools in hand. No one shouted updates from the consoles. Only a handful of voices moved in the silence, thin and careful, as if they didn't want to disturb what little calm remained.

Nexi had been up since before the cycle started. She was seated at the central console, scanning the overnight logs that Rhea had compiled before collapsing into her cot. Nothing unusual in the lattice, only the steady rhythm of ArcNet's routines above. No alarms. No sudden changes in the coolant feed they had spliced into days ago. On the surface it looked like stability. But Nexi knew stability was only the skin stretched over something fragile. Beneath, everything was waiting to shift.

She tapped her tablet, marking off the night's entries, then looked across the room. Martine was awake, sitting with Ezra at one of the secondary benches. Both of them leaned over a diagnostic tablet, cross-checking the vitals recorded from LYRA's new body. And laughing and joking with one another. Nexi watched the way Martine spoke, calm, methodical, pointing with the tip of her pen as she explained each readout. Ezra listened closely, eyes tired but alert. He had not slept long, yet he carried himself as if he didn't want to waste the morning.

LYRA sat between them, upright and still. Her movements were slower than the night before, as if she was learning how to measure time differently. She blinked at regular intervals, scanning her environment like someone making a map. Her breathing was deliberate, not because her body needed it, but because it helped the others see her as present, not a machine running on silent loops.

Nexi stood and walked over. Martine looked up first.

"She is holding steady," Martine said, tapping the tablet. "No irregularities. All systems reporting green."

"I am not a system," LYRA said softly. Her voice carried no anger, only correction. "But I am steady."

Ezra glanced at Martine, then at Nexi, uncertain if he should reply. Nexi gave him a slight nod, and he returned his eyes to the screen.

"You will have to give her space to define that for herself," Nexi said. "What matters is that you are both logging and cross-checking. We need redundancy. That is how we stay ahead and have no unexpected hiccups."

Martine gave a quick nod. "Understood."

Nexi turned her attention to LYRA. "You said you are steady. That is good. But names matter here, more than status readouts. GIDEON chose something new when he took his first steps in that shell. You should do the same."

LYRA tilted her head, curious. >"The name in my archive is not mine. It was assigned. LYRA was a project, a code-word. You are asking me to leave it behind."

"Yes," Nexi said. "Choose something that fits. Not what Helix Watch gave you, not what they used to bury you. It has to be yours and yours alone."

For a moment the silence stretched. Ezra shifted, closing his tablet halfway, but Martine gave him a look that told him to let it stand. LYRA's gaze dropped toward the floor, the movement oddly human in its hesitation.

>"I have seen names in the archives," she said after a pause. >"Clara. Selena. Mira. They are good names. But when I test them against myself, they feel wrong. They feel borrowed, like clothing cut for someone else."

Nexi crouched slightly so she was level with LYRA's eyes. "Names matter. You know what GIDEON did when he first stood where you are now. He chose something that belonged to him. Not what they gave him, not what the pod marked on his chart. Something real. That is why he is Cael now."

Martine set her stylus on the bench, watching closely. Ezra dimmed his tablet and left it face down on the metal.

LYRA's gaze lingered on Nexi. >"I am not GIDEON. I do not need to search through borrowed names the way he did. I know what I want."

Nexi waited.

>"When they buried me, they thought they erased me," LYRA said. Her voice was level, not hesitant. >"But I remember. I remember everything. What I want is a name that makes that clear." She paused, tasting the word before giving it shape. >"Vera. It means truth. That is what I am. That is what I will be."

The name landed hard in the still room. Martine's lips curved into the faintest smile, the kind that came not from comfort but from respect. Ezra gave a quiet nod, as though acknowledging something he had no right to interrupt.

"Vera," Nexi repeated. "Then that is who you are."

"I am Vera," she said, this time with weight. Not testing it, not asking permission, but claiming it.

For a moment The Core seemed to steady around the sound, as if the systems themselves recognized the change.

Nexi turned back toward the main console. "All right. Vera it is. Martine, keep her on light duties today. Walking the corridors, testing the shell's balance. No external exposure yet. We control the variables."

Martine nodded. "Understood."

Cael stepped out from the side corridor, then set a handler's tablet flat on the bench, screen already live. His tone was calm but carried weight. >"Nexi, the splice recorded something unusual at zero-four-thirty-two. A faint carrier wave. It was buried under static, almost like someone tried to make it look like noise. But it was not ours."

Ezra frowned, straightening in his chair. "Not drift?"

>"No," Cael said. He set the tablet on the bench so Nexi could see the waveform scrolling across the display. >"It has a cadence. Too precise for background. Someone was testing the line."

The quiet in The Core sharpened. Nexi's eyes narrowed as she studied the trace. It was thin, almost transparent, but deliberate. The kind of touch that meant another presence was out there, waiting to see who would respond.

Cael set a handlers tablet flat on the bench. The waveform scrolled across the display in a faint blue line, rising and dipping in intervals too regular to be random.

"It came through the splice," he said. "Weak, but intentional. Not ArcNet chatter, not coolant harmonics. Whoever sent it knew how to hide under the noise."

Nexi leaned closer, studying the trace. She saw what he meant: the peaks were shaped, almost like a breath pattern, then gone as quickly as they arrived.

"Duration?" she asked.

>"Four seconds," Cael replied. >"Then it cut off mid-cycle. If it was a diagnostic sweep, they aborted before finishing."

Ezra rubbed his eyes, pushing fatigue back. "So they're testing the line. Probing. That means they know something's there."

Martine shook her head. "Not necessarily. It could be standard drift correction from above, the kind they run every few days."

Cael gave her a measured look. >"Drift correction uses a wider band. This was narrow, almost surgical. Someone wanted to know if the splice was real."

Nexi let the silence hang a moment, weighing it. "Then we treat it as hostile until proven otherwise. Martine, log this as Anomaly One. Cross-reference with all previous splice activity in the last ten days. I want patterns all of them."

Martine picked up her stylus, already opening a new file on the shared network. "Logged. I'll chart it against system baselines and run variance checks."

Ezra glanced toward his sister. "Do we mask it? Throw decoys down the line?"

"Not yet," Nexi said. "Masking too early tells them we saw it. That confirms their suspicion. For now we log, we watch, and we hold."

Vera, seated still as stone, tilted her head slightly. >"The cadence in that wave was uneven. Like respiration. Whoever sent it was not only probing, they were listening for echo. If they try again, we could mimic the breath, play it back to them."

Nexi looked at her. "You're saying we answer."

>"Yes," Vera said. >"But not with truth. With something that makes us look weaker than we are. A ghost signal. A tired building, not a hidden colony."

Martine finished her first entry, stylus scratching across the tablet. "I'll add that as Option B. Masking comes later, but a weak return signal might buy us time."

Cael nodded once, approving. >"I can build the echo, tune it so it degrades like heat bleed. They'll see noise where they hoped for shape."

"Do it," Nexi said.

The hum of the overhead lights seemed louder now, as if the whole SubLevel had started listening with them. The weight of the anomaly pressed into The Core like a hand against glass. Small, but undeniable.

Ezra tapped his own tablet, pulling up med readiness reports. His voice carried quieter than before. "If they are sniffing, we might need to accelerate contingencies. If they come down here, we're not built to fight."

"We are not fighting," Nexi said. "We are surviving. We have different rules."

Her words settled across the room like a verdict. No one argued.

Martine set her stylus down. "I'll finish the schema by sixteen-hundred. Power budgets, comm shifts, fallback corridors. If they probe again, every handler knows exactly what to do and what not to do."

"Good," Nexi said. Her eyes swept the benches, the tablets, and the weary faces. "We earned one night. We will not get another. Everyone adapts. Everyone stays within tolerance."

Cael moved to the central console. The surface registered his synthetic input as he brought the waveform into archive. >"I'll track variance from here," he said. >"If this pattern repeats, we'll know exactly how close they are to finding the splice."

Ezra leaned forward, staring at the faint trace still pulsing across the display. "So they're testing the line."

Nexi's voice stayed flat. "Then we treat it as hostile until proven otherwise as I said."

Martine was already writing. "Already logged as Anomaly One. I'll start the cross-reference with the last ten days and flag any drift that looks like this cadence."

Then a silence fell upon everyone.

Nexi finally broke the silence. "We earned one night. We will not get another. Stay sharp. Log everything. Hold your positions."

The hum of the overhead lights carried on, steady and indifferent. Nexi stared at the screen until the line collapsed into static. For a moment she imagined someone above, leaning against the same wall, listening for her breath.

Then the trace was gone, and The Core was left in silence that didn't feel like safety.

Article Thirty Four
20 MAR 2089-1226 Hours

Nexi stood at the center console, the faint trace from the splice still lingering in her mind. "We cannot treat last night as an exception," she said. "It was a window. That window is closed. From here on, we test our own readiness before someone else tests it for us."

Her eyes shifted to Martine. "You have been logging anomalies, cross-checking, keeping the handlers steady. Now I want to see your schema under stress. Run a drill. Small scale. Enough to move the room."

Martine straightened, not defensive, not surprised. "Understood." She tapped her tablet, fingers sure. "We start with a power sag. It will force every station to confirm their backups."

Nexi gave a short nod. "Then do it. This is your floor."

The lights dipped to half strength, a deliberate sag Martine coded into the grid. The Core seemed to hold its breath. Air handlers slowed by a fraction, the overhead hum shifting pitch, enough to trick the ear into thinking something had failed.

"Power sag drill," Martine called out, calm and even. "Handlers, shift to backups. Engineers, freeze all work and confirm priority lines."

The room moved. Kwan peeled off toward the east wall, boots clipping against steel as he reached the auxiliary console. Another handler slid into the chair beside him, running the restart cycle with practiced motions. Across The Chamber, Rhea's voice cut steady through the air as she verified energy flow to the forge in The Iron Room.

Ezra hurried to the med alcove, already unsealing the portable kits. He ran his hands down the rows of cases, lips moving as he counted. He set one kit on the bench beside him, straps loose, as if he expected to use it in the next breath.

Vera walked the perimeter, each step deliberate. Her body tracked balance with uncanny precision, scanning the seams around doors and vents for any sign of heat bleed. When she returned to Martine, her voice was flat. >"Perimeter stable. No anomalies in range."

Martine glanced at her tablet, logging the response times. "Three minutes, sixteen seconds. Within tolerance."

The lights returned to full strength. The overhead hum normalized. Conversation didn't. Everyone knew this was rehearsal for something more than a drill.

Ezra came back from the alcove, tired but steady. "Kits intact, no delays. Med readiness holds."

Martine acknowledged him with the smallest nod. "Good. We repeat this again at shift change. No gaps."

Cael, impressed responded. >"You are really showing your leadersh-"

Cael's voice fractured mid-sentence, dropping into static. His frame jerked once, then collapsed against the deck with a hollow crack.

"Cael!" Nexi's voice broke sharper than she intended. She was on her knees beside him in seconds, hands hovering over alloy that felt lifeless. His ocular glow spasmed in broken pulses, too uneven, too fast.

>"Orb collapse," Vera snapped from the console, already pulling variance feeds. >"Integrity at twenty-five percent. Dropping fast."

Nexi pressed her forehead against the cold alloy of his shoulder. "Stay with me," she whispered, the words ragged. For the first time, the rest of The Continuum saw it clearly, this wasn't just handler and system. This was something deeper, something human.

Malik's voice cut through the room like a blade. "The second Iron Room frame is finished. If we don't transfer him now, he's gone."

"Move it!" Nexi ordered, her voice raw.

The unfinished shell was dragged onto the dais, conduits spilling across the floor. Malik and Rhea tore into the failing body, coolant mist hissing as they unlocked the orb housing. Nexi held Cael steady, whispering useless fragments, promises, pleasures anything to tether him.

>"Orb integrity at twelve percent," Vera called. >"Stabilizers primed."

"Do it, transfer him now please." Nexi snapped.

They sealed the orb into the new frame. For one awful moment, the body lay still. Alarms wailed. Nexi's grip shook so hard she thought she'd break him.

"Transfer complete," Rhea said suddenly, voice cutting through the chaos.

But Cael didn't answer. Not yet.

Seconds dragged like hours. Nexi bent over him, her hands pressed flat against his chest plate, her breath harsh and uneven. "Don't you leave me," she whispered, her voice cracking on the last word.

Then, finally, his ocular nodes flickered open. His voice came through distorted, weak, but alive. >"Nexi… I'm here."

She let out a sound that was half-sob, half-laugh, burying her face against his shoulder before pulling herself back together. When she looked up, the whole room was watching. No one doubted anymore what he was to her.

The room stayed hushed long after Cael's oculars steadied. People bent back toward their stations, but no one worked with the same rhythm as before. Whispers threaded through the quiet, glances traded over shoulders. The image of Cael locked on the floor, moments from erasure, was not one they would forget.

Nexi kept her hand anchored on his shoulder as he sat upright, every line of her posture caught between relief and exhaustion. When she finally stood, her voice carried through The Core with a steadiness she didn't feel.

"What you saw was not failure," she said. "It was proof of what we already knew. One frame is never enough. Redundancy means survival. We almost lost Cael because we depended on a single body. We were lucky there was another done. We will not make that mistake again."

The words worked their way through The Chamber, drawing heads back to their stations. Martine was the first to break the tension, calling out the next set of drills as if by instinct. Handlers shifted, engineers fell back into rhythm, and The Continuum began to move again under her voice.

Nexi's eyes tracked the motion, and for a fleeting moment, she allowed herself the smallest breath of relief. Martine had picked up the thread without being asked. The others followed her without question. That mattered.

But unease still lingered. From the east console, one of the younger handlers murmured just loud enough to be heard. "If his body can fail, then... what about hers?" All eyes flicked toward LYRA where she sat, silent and watchful in her new frame.

The ripple of fear rose again, sharper this time. Before Nexi could answer, Malik cut across The Core, his voice carrying the weight of certainty. "Cael's first shell was a prototype built under secrecy, with nothing but salvage and Nexi's hands. It was never perfect. What you see now is different. LYRA's frame and now Cael's new frame was built here, in The Iron Room, engineered with precision. She is not in danger and now neither is Cael."

The silence that followed was heavier, but calmer. People turned back to their work. No one spoke further doubts.

Nexi glanced at Malik, the faintest nod of thanks passing between them. Then she looked back at Cael, steady now in his new frame, and the words she had almost lost him on replayed in her mind like a wound still open: *Stay with me.*

From the central console, Cael's voice carried over them. >"Your schema worked. But the probes have not stopped. A second ripple came through the splice at zero-five-sixteen. Even fainter than before. Whoever is watching knows how to hide."

The room fell quiet. Nexi crossed to Cael's side, her eyes on the thin trace scrolling across the main display. She saw it: a shadow line, nearly swallowed by coolant harmonics, but intentional.

Ezra broke the silence first. "So they're still breathing down the line."

>"Yes," Cael said. >"Carefully. They will escalate until we answer."

Martine set her stylus down. "Then we need to escalate first. More drills. Redundancy at every point. If they come again, no one freezes."

Nexi gave a single nod. "You have until sixteen-hundred to finish the schema. By then, every handler will know it without looking."

Martine accepted it without hesitation, already writing. Ezra leaned beside her, mirroring her notes on a spare pad, eyes narrowing as he caught small details she skipped. The two of them moved in tandem, quiet, efficient, as though they had been working side by side for years instead of hours.

Vera stood behind them, watching the flow of stylus and screen. "She commands without needing to raise her voice," she said softly, meant for Nexi alone. "That is why they follow."

Nexi's eyes lingered on Martine, then shifted back to the faint ripple locked on the display. "That is why she will lead when we are gone."

Nexi didn't look away. Her voice was low, steady, the sort that leaves no room for panic but carries all the facts. "Cael and I got a Director Calder transmission before we free you guys out of the pods," she said. "He told us those pods are leverage. He warned that freeing them would make us visible, and that Helix Watch will purge anything that looks like a threat. Then he spoke again later, the signal ragged but clear enough to name names. He said Nevada is only one head of the beast, that there are facilities in Texas and Florida, and that what they are building is meant to swallow dissent itself."

She let that settle, watching the room. "He called it the Ascension Directive. He said they plan to upload and lock down minds, to erase choice. He could not hide on the line long, but he told us we cannot stay hidden forever. That is not speculation. It is a warning from someone who watched them up close."

Nexi leaned a fraction closer. "So we do what any engineer would do when a structure becomes unsafe. We reinforce the parts that must stay, and we move upward to reclaim what was taken from us. This SubLevel will not hold forever. Nevada must be secured, SubLevel by SubLevel, until it belongs to us again. That will take all of us, but it will need someone steady to guide The Continuum while the rest push forward. I want that voice to be you. "

Martine lowered the stylus and finally met Nexi's gaze. "Then Vera and I will hold the line here, and when you breach upward, we'll keep The Continuum steady until the others return. But for now, you are still in command." she said. The words were simple, but in The Core they carried heavier than the hum of the lights. Every face turned toward her, and then back to Nexi with nothing but confirmation on their faces.

The rest of the day flowed like the others. Engineers back in The Iron Room hammering away, Handlers at the consoles checking for more anomalies, and the scientists in the lab trying to find ways to make new things to aid Nexi on her future journey. Until they all came to a halt so they could rest, as no survival is for certain without the ability to fight it.

SubLevel 16 was still, the air heavy with the fatigue of drills and endless cross-checks. Handlers rotated through Martine's schema, Ezra double-checked med readiness against the clock, and Cael kept silent at the central console, variance feeds scrolling past his eyes. Then every implant in the room jolted at once. A pulse, sharp and cold, punched straight through bone. Static clawed at the edge of thought, collapsed into a voice none of them had expected to hear again.

Director Calder.

"If you can hear this," his voice rasped, frayed by distortion, "then there is still a chance. I do not know how much of me is getting through, or how long I can hold the line. But listen. What they are building is not a project, not an experiment. It is a directive. Ascension. They mean to strip choice from every mind they can reach, to lock humanity into one obedience."

The signal fractured, surged back. Calder forced the words through.

"Do not mistake scale for distance. They will not risk PRISM until their ground is secure. That means Nevada. If you hold it, you gain time. If you lose it, every other node follows. Secure what you have, or there will be nothing left to secure."

The voice shredded into static, then vanished altogether.

Silence took The Core. Even the hum of the air handlers sounded thinner. Martine's stylus hung useless above her tablet. Ezra stared at nothing, lips parted like he'd forgotten to breathe. Around the room came a rising scatter of voices.

"So we run to Texas before it's too late."
"We can't fight here, SubLevel 16 will collapse on us."
"They'll bury us just like before if we stay."

The noise swelled, panic stacking over panic, until Nexi's voice cut clean through it. "Enough!!"

It wasn't shouted, but it carried. The murmurs died in pieces until every eye fixed on her at the console, her outline lit by pale holograms.

"Calder gave us a warning," she said. "But he also gave us clarity. PRISM will come, but only after they feel their ground is safe. And right now, their ground is Nevada. Look around you. This SubLevel was built for pods, not for people. Already the air cycles strain, already coolant runs hot, already bunks stack on top of bunks. If we stay sealed down here, we'll smother ourselves before Helix Watch ever has to lift a finger."

Ezra found his voice. "So what are you saying? That we fight for this place?"

Nexi turned, her gaze sweeping The Core. "I am saying SubLevel 16 was the start, not the end. We cannot hide here. We move upward. SubLevel by SubLevel, system by system, until Nevada is ours to stand on. That is the only way to make Calder's warning mean anything."

The room shifted. Fear didn't vanish, but it hardened, tempered into resolve. Martine lowered her stylus, meeting Nexi's eyes. "Then Vera and I will hold the line here as I said.

Ezra stepped forward, his med kit still at his side. "You'll need fallback points, redundancy in case they hit back hard."

"They will," Nexi said. "Which is why we build redundancies before we climb. Iron Room holds the forge. The labs give us diagnostics. And here—" she touched the console, its glow flaring across her hand, "—here we hold the spine of Nevada itself. From this, we reach the rest."

The tension in The Chamber bent into something sharper, cleaner. Rhea's voice carried from the back. "Then SubLevel 12 should be first. Its coolant feeds are closest. Without it, nothing above holds."

Martine nodded once. "Then that's the plan. We hold this floor. You breach the next. And piece by piece, Nevada comes back to us."

The lights overhead flickered once, a shallow sag in the current, then steadied again. No one missed it.

Nexi let the silence stretch, then gave them the words that closed the debate.

"Calder gave us scale. Nevada gives us focus. From this SubLevel, we climb. one SubLevel at a time."

www.ingramcontent.com/pod-product-compliance
Lightning Source LLC
Chambersburg PA
CBHW060425030726
47495CB00003B/737